A GUIDE TO METRICATION

BY

M. J. B. JONES

THE QUEEN'S AWARD TO INDUSTRY 1966

PERGAMON PRESS

PERGAMON PRESS LTD.

OXFORD · LONDON · EDINBURGH
NEW YORK · TORONTO · SYDNEY

First Edition 1969

Library of Congress Catalog Card No. 79–82383

Printed in Great Britain by A. Wheaton & Co., Exeter

08 006539 2

Contents

	FOREWORD, *by* SIR ANTHONY BOWLBY, BT.	vii
	PUBLISHER'S FOREWORD	ix
	PREFACE	xi
	ACKNOWLEDGEMENTS	xiii
1.	*The Case for Metrication*	1
2.	*The Origin and Development of the Metric System*	5
3.	*System International Units*	15
4.	*National and International Standards*	35
5.	*Additional Benefits of Metrication*	42
6.	*Precise Conversion of Imperial to Metric (SI) Quantities*	50
7.	*International Standard Drawing Office Practice*	57
8.	*The ISO System of Limits and Fits*	64
9.	*The ISO Metric Screw-thread System*	76
10.	*Currency Decimalization*	89
11.	*Metrication and Modular Construction*	106
12.	*Metrication in the Electrical Industry*	110
13.	*Metrication in Mechanical Engineering*	114
14.	*Education and Training*	121
15.	*The Preparation of Changeover Programmes*	126
	BIBLIOGRAPHY	136
	APPENDIX *1.* *Conversion Factors*	140
	APPENDIX *2.* *Published Metrication Programmes*	146
	INDEX	153

Foreword

SIR ANTHONY BOWLBY, Bt.

Chairman, G.K.N. Screws and Fasteners Ltd.

Past President and Chairman, British Standards Institution

IN MAY 1965 the Board of Trade issued a policy statement which made it clear that the Government were determined to implement the adoption of the metric system. Any change raises the question of justifying the costs by the advantages derived, but one as fundamental as replacing accepted Imperial units with metric equivalents is bound to raise many questions.

This book illustrates the inevitability of the change and on this assumption there must be an optimum point in the future for each industry to convert to the metric system. The period of time before replacing all Imperial terms must be long enough to fully plan the education and re-equipment programme, but delay beyond this point can only involve still greater costs.

The facts presented give a full insight into the definitions of basic and derived units, the standards documentation and conversion factors. The information is presented in a logical manner which will be of particular interest to all who will become intimately concerned with this subject in the next decade.

Publisher's Foreword

THE Industrial Training Act has resulted in an increase in the number of people now being trained or re-trained. LICET books are intended to provide suitable texts which will be easy to read and assimilate for those employed in industry and commerce who are receiving further education and training as a part of their employment. It is hoped that they will be particularly suitable for those attending courses leading to the examinations of the City and Guilds of London Institute, the Regional Examining Unions and other examining bodies.

The books are essentially straightforward, simple and practical in their approach and are designed to provide all the basic knowledge required for a particular trade or occupation. They are structured in such a way that the subject is broken down into convenient and progressive components, and are written by authors specially chosen for their expert knowledge and for their practical and teaching experience of their subjects.

Where appropriate, emphasis has been placed on safety training. In some subjects separate manuals on safety and safety training will be provided; in other texts, authors have been encouraged to emphasize safety precautions at relevant points, or to devote a separate chapter to these matters.

LICET books are published in a number of subject divisions, with each division controlled by a specialist editor responsible for selecting authors and providing guidance and advice to both authors and publisher. It is hoped that the series will make an important contribution to further education and industrial tra ining.

ROBERT MAXWELL
Publisher

Preface

THE structure of a modern industrial society contains an intricate network of interrelationships between various sectors of industry, commerce, government and education. In considering the fundamental changes involved in introducing the new primary metric system of weights and measures the first problem is where to begin. In so far as manufacturing industry is concerned it is clearly necessary for that sector which produces components for assembly industries, to offer components to the new dimensions and standards before the major changeover can begin.

The material in this book was assembled in the course of preparing retraining lectures for staff employed in the G.K.N. Group Technological Centre, and in writing reports for the guidance of various departments of Group Companies such as design, marketing, material testing and so on.

The justification for the presentation of the material in book form is that, in order to obtain the information in any other way, one would have to purchase a large number of separate publications, and read far more print than is presented here.

It is hoped that the book will be of value to the smaller industrial organizations which are unable to justify the setting up of an information unit or standards department. There are many such organizations and the success of the metrication programme depends upon the concerted action of all. Without good communications this end cannot be attained.

The book is published by permission of the Corporate Staff Director of Research and Products, Dr. T. Emmerson.

My activities throughout the metrication exercise have been under the immediate supervision of Mr. H. K. Moneypenny, whose advice and constructive criticism is gratefully acknowledged. Many members of the G.K.N. Group Technological Centre Staff have contributed advice and assistance, for which I here record my thanks.

In accordance with the recommendations of the Decimal Currency Board, with which the Ministry of Technology and the British Standards Institution have agreed to comply, the decimal marker used throughout is the full point.

Large numbers may be divided into groups of three digits either by a comma, or by a space.

The Decimal Currency Board use the comma for thousands, there being a marked reluctance in the financial world to leave spaces in rows of figures.

The British Standards Institution prefer the use of a space for thousands. The reason for the preference is that when engineering drawings and technical literature circulate between the United Kingdom and Europe there is a risk of confusion between the European use of the comma as the decimal marker and the U.K. use of the comma for marking off thousands.

Thus, when writing one thousand two hundred and thirty four decimal five one could write

1,234.5	Decimal Currency Board Style
1 234.5	British Standards Style.

M. J. B. Jones

Acknowledgements

MY THANKS are due to the following individuals and organizations for permission to publish copyright material:

British Standards Institution, for material from the British Standards listed in the bibliography, and for the use of tables and illustrations from BS Handbook 18: 1966, *Metric Standards for Engineering*:

Tables 5.1, 5.2.

Figures: 8.1, 8.2, 8.3, 8.4, 8.5, 8.6, 8.7, 9.1, 9.3, 9.4, 9.5, 9.6, and Fig. 15.1 from the BSI publication *The Adoption of the Metric System in Engineering, Basic Programme and Guide.*

BSI publications are available from:

The Sales Branch,
101–113 Pentonville Road,
London N.1.

Industrial Research Inc., of Beverly Shores, Indiana, U.S.A., for the results of their poll, quoted in Chapter 2, of opinion in American industry regarding metrication.

The Editor, the *Engineer*, and Mr. G. E. Dodd of English Electric Co. Ltd. for Fig. 9.8 which was published in the *Engineer* of 16th February 1968, on page 279, in correspondence on screw thread standards.

The Director and Secretary, the Engineering Equipment Users Association, for their programme 'Change to the Metric System'.

The Secretary, the Institution of Production Engineers, for the programme published in the booklet *Metrication in the United Kingdom.*

The Education and Training Officer, Central Electricity Generating Board, for the CEGB Notional Plan for Metrication.

The Editorial Director, John Wiley & Sons Ltd., for Fig. 3.1, Viscosity Conversion Chart, reproduced from *Theory and Practice of Lubrication for Engineers*, by Dudley D. Fuller.

The Chief of Special Services, *The Financial Times*, for the list of official exchange rates for world currencies.

Materials Management Institute, Boston, Mass., U.S.A., for material from PERT and CPM by David M. Stires and Maurice M. Murphy.

Mr. G. T. Hughes, of G.K.N. Head Office, for clearance on legal matters.

Mr. R. Kelsall, Head of G.K.N. Group Public Relations, for clearance of the text on publicity matters.

Information on government policy in such matters as Education and Training was obtained from Crown Copyright Documents.

The section in Chapter 15 on Metrication in India owes much to the report prepared for BSI by Mr. G. E. Dodd and Mr. G. T. Pearmain.

Mr. H. A. R. Binney, Director of the British Standards Institution, for arranging a final scrutiny of the manuscript by BSI, and Col. J. S. Vickers for carrying out the scrutiny and offering valuable constructive criticism while under great pressure of work.

M. J. B. Jones

CHAPTER 1

The Case for Metrication

1.1. The Technical Case

In one sense it is true to say that the movement towards Britain's adoption of metric weights and measures began in 1864. In that year Parliament legalized the use of metric units in contracts. From that time onward Britain has been bilingual in weights and measures, Imperial and metric units appearing side by side in everyday contexts. It has therefore been possible to compare the two systems in use over a long period.

The most puzzling feature of the Imperial system is the variety of multiples: 12 inches to a foot, 3 feet to a yard, 22 yards to a chain, 80 chains to a mile being sufficient to illustrate the point. Historical investigation has revealed that the reason for this variety of multiples is that Britain underwent a series of invasions, culminating in the Norman Conquest. Each successive wave of invaders introduced their own units and crudely related them to the indigenous system by minimum adjustments. The Imperial system thus grew into a collection of unrelated units bearing names derived from several languages. The unrelated sequence of multiples and the fact that the names of the units bear no relation to their magnitude, or to their relationship with other units, makes the Imperial system cumbersome to use in conjunction with modern calculating aids.

The metric system is, by contrast, of recent origin, and is based on a single multiplier, namely ten. The decimal structure of the system makes the design and construction of calculating aids relatively straightforward.

1.2. The International Compatibility Case

The twentieth century has been a period of rapid growth of world usage of the metric system, to the extent that by 1967, 85% of the world's population were living in countries which made the metric system their primary system of weights and measures.

Having an economy which is dependent upon exports it is inevitable that Britain should pay attention to the practice of other nations in all matters which could affect the acceptability of her products in export markets. Once the fact was established that over half Britain's exports were to metric nations a case could be made out for the adoption of the metric system which was quite separate from any arguments based on technical merit. The argument that it is advantageous to use the same basic units of measurement as the majority of mankind is, of itself, a sound one. There is, however, an extension to it which further strengthens the case for metrication.

There exists a range of standards and codes of practice, expressed in metric terms, and promulgated by the International Standards Organization. The adoption of these standards will ensure the acceptability of Britain's exports in the sixty nations which are members of the International Standards Organization, and to a considerable extent, in those nations which are metric but not International Standards Organization members. One of the advantages of adopting ISO practice will be that reductions can be made in the variety and range of materials, components, and products which have to be produced.

The capital tied up in stocks of material and components depends upon two factors: one is the variety required, and the other is the quantity of each item required. The variety factor is the predominant one in determining the cost of stockholding. When all suppliers and users base their specifications on a single set of standards, expressed in terms of one system of units, significant economies can be made by variety reduction.

In connection with the adoption of standards it is important to realize the difference between metrication and pseudo-metrication. The translation of the dimensions of an article from inches into millimetres is pseudo-metrication. A 4-inch pipe remains a 4-inch pipe even if it is called a 101.6-mm pipe. A metric pipe of comparable size would have a diameter of 100 mm. The module must change before the claim to have 'gone metric' can be upheld.

There is, to the above argument, an interesting corollary: International Standardization proves to be a two-way traffic between the metric and Imperial worlds and it is worth enquiring which inch-based engineering standards are acceptable in the metric world. When these are known, the question to be considered is whether or not there is any useful purpose to be served by changing the standards. To quote two simple examples, and there are many more, printed circuits for electronics originated in the U.S.A., and the dimensional standards relating to them are based on a module of one-tenth of an inch. Printed circuits made and used in continental Europe are usually to these standards. There seems to be no dissatisfaction with the standards and no great desire to change them. Large-diameter pipes, as used in the oil industry, are based on inch standards. In this case also the industry is largely American based. The inch standards are known, understood and accepted in the metric world.

1.3. The Abandonment of Imperial Measure

Faced with the two facts that the Imperial system has inbuilt disadvantages and that it is now a minority system, British industrialists have exerted pressure on the Government to support their policy of metrication. The existence of a strong body of opinion became obvious in 1951 when the Hodgson Committee, sponsored by the Board of Trade, reported. The Hodgson Committee Reports, however, added the condition that the change should be made in concert with the other major inch-using nations, notably America and Canada. Since there was, at that time, no significant lobby for metrication in America, no action was taken. Since 1951 the pressure has been applied through the British Standards Institution and the Confederation of British Industry. In May 1965 the industrialists' policy of metrication in manufacturing industry received the support of Government.

The proposed national plan is to embark on the systematic changeover of industry in 1970 and to complete the bulk of the work by 1975. The official announcement of the national plan is contained in the British Standards Institution Document PD 6424, July 1968, *The Adoption of the Metric System in Engineering: Basic Programme and Guide.*

CHAPTER 2

The Origin and Development of the Metric System

2.1. The Unit of Length

After the French Revolution, in 1789 to 1792, the new governors of France were faced with a chaotic situation arising from the use of different standards of weight and measure in different areas of the country. The French attempted to define the metre in terms of a geodetic constant. The metre was first defined as one ten-millionth part of the Earth's Polar Quadrant passing through Paris. The practical embodiment of this defined length was a platinum end standard called the 'Mètre des Archives'. The geodetic definition unfortunately proved to be inexact and the Mètre des Archives had to be established as the primary length standard in France in 1799. A platinum–iridium International prototype metre was approved by the International Conference on Weights and Measures in 1889.

In order to avoid the difficulties associated with bar standards, a suggestion made by the American physicist, A. A. Michelson, was eventually adopted. Michelson, who was awarded a Nobel Prize in 1907, studied spectroscopy and metrology and, as a result of his studies, suggested that material length standards could be replaced by reference to the wavelength of light. In 1892–3 Michelson measured the metre in terms of the wavelength of cadmium light and since that time basic units have, where possible, been referred to natural phenomena.

The Eleventh General Conference of Weights and Measures in

1960 adopted, as the definition of the metre,

1,650,763.73

wavelengths *in vacuo* of the radiation corresponding to the transition between the energy levels

$2p_{10}$ and $5d_5$ of the krypton-86 atom.

Since January 1964 the Imperial Standard Yard has been defined as

$$1 \text{ yd} = 0.914\,4 \text{ metre}$$

which is equivalent to 1 inch = 25.4 mm exactly.

At the same time the Imperial Standard Pound was defined in terms of the kilogramme and the basic standards of the customary system of Weights and Measures of the United States of America were brought into line with the Imperial System by redefinition in identical terms. There is thus, at present, one basic set of primary standards in the world, and those are the metric standards.

2.2. The Unit of Mass

The first material embodiment of the unit of mass was the kilogramme des archives. This was a solid cylinder of platinum which served as the primary standard throughout the nineteenth century. In 1889, however, a new standard was constructed of platinum–iridium (90% Pt, 10% Ir). The international prototype kilogramme, as it was called, was a cylinder of height equal to its diameter. The greatest source of error with a primary standard of mass is abrasion due to handling. Care has to be taken to keep the number of references to the primary standard to the absolute minimum and to use apparatus and techniques to minimize the risk of abrasion.

In carrying out standardizing measurements of mass, which are made indirectly by weighing, the buoyancy effect of the displaced air is taken into account. Double weighing is carried out as a precaution against asymmetry in the balance, and careful temperature control is maintained. The efficacy of the techniques employed, and the stability of the primary standard and its copies, are well demonstrated by the record of comparisons between the primary standard kilogramme and the British Copy Number 18. Over a period of seventy years, and five comparisons, the variation has been better than one part in 10^8.

The kilogramme was originally linked to the unit of length by defining it in terms of a volume (that is: length^3) of water at 4° Celsius, which is the temperature at which water is at its maximum density. This attempted linking of the basic units caused some confusion owing to slight inaccuracies in the original measurements. The kilogramme is now defined as the mass of the international prototype kilogramme at Sèvres.

Since January 1964 the Imperial Standard Pound has been defined in terms of the kilogramme and is defined as

$$1 \text{ lb} = 0.453\ 592\ 37 \text{ kg}.$$

In this sense the British standard is a metric standard.

2.3. The Unit of Time

Until the atomic clock was devised and constructed, in 1955, the second had to be derived from astronomical observations. Because the Earth's rate of rotation is not constant, and because the Earth wobbles slightly on its axis, the universal second obtained from observations of the transit of stars varies by about one part in 10^8 over a period of one year and shows an average increase of one part in 10^8 over a century. In order to obtain a more stable standard a fictitious unit called the ephemeris second was devised. This is defined in such a way that it is equal to the mean value of the universal second during 1900.

The difficulties associated with attempts to obtain a stable basis for the definition of the second led to consideration of atomic energy transitions as a basis of time measurement. The transition between two energy states of an atom is accompanied by the emission or absorption of radiation. The frequency of the radiation is a constant depending only on the structure of the atom. This statement is true so long as the energy levels involved are not being influenced by external magnetic fields. Since frequency is the inverse of time interval, the frequency of the radiation emitted provides a unit of time interval.

It is important to distinguish the idea of time interval from that of time in the astronomical sense, which is a measure of the Earth's rotation. Since 1964 an interim definition of the second as a unit of time interval in atomic terms has been in use. The second is defined as the interval occupied by

9,192,631,770 cycles of the caesium transition.

2.4. The Development of the Metric System

From the original founding of the basic units the metric system has undergone three major reconstructions. From the original system came the centimetre-gramme-second or cgs system familiar to all those trained in the natural sciences. The late nineteenth and early twentieth century was a period of rapid growth of the natural sciences, in particular the theory of electromagnetism. Attempts to understand the phenomena of electricity and magnetism led to the use of three sets of units: cgs electromagnetic, cgs electrostatic and 'practical'. The confusion caused by this three-compartment treatment of what is really one body of knowledge led Professor Giorgi, of Rome, to propose a fresh approach to the whole question of units.

In 1904 Professor Giorgi proposed that the basic units should be the metre, the kilogramme and the second. To these three basic units a fourth, electrical, unit should be added. The choice of the electrical basic unit is arbitrary and, eventually, the ampere was chosen. The ampere was defined in terms of the force between conductors in each of which a current of one ampere was flowing. Giorgi showed that a system so based would include the practical electrical units, the volt and the ohm, as natural, absolute units. The MKSA, or Giorgi, System of Units was formally adopted by the International Electrotechnical Commission in 1938.

By 1960 agreement had been reached among the member nations of the International Standards Organization on the addition of two further basic units to the MKSA system. The degree Kelvin was introduced as the the unit for measurement of thermodynamic temperature, and the candela was introduced as the unit for measurement of luminous intensity. It was further agreed that a refinement of the electrical unit system, known as 'rationalization', should be carried out.

Rationalization has the effect of giving electrical formulae a more logical appearance by ensuring that the factor 2π appears only in those cases where cylindrical symmetry is involved, and the factor 4π appears where there is spherical symmetry. To attain this end it is necessary to assign a value incorporating the factor 4π to the constant known as the 'magnetic space constant' or the 'permeability of free space', symbol μ_0. SI assigns the value $4\pi \times 10^{-7}$ to μ_0. In 1960 it was agreed that the rationalized and supplemented MKSA system should be named the System International, abbreviation SI, in all languages.

2.5. The Growth of Metric Usage

Concurrent with this evolution of the metric system was the gradual spread of metric units as an international language of weights and measures. Figure 2.1 shows the cumulative total of the number of nations of the world using metric units, plotted against time. It is instructive to consider some key dates and to relate these to Fig. 2.1:

1795: The metre became the legal basis of linear measure in the third year of the French Republic.
1834: All other units were made illegal in France. This step was taken after a new generation had become familiar with the metric system in school.
1864: The English Parliament legalized the use of metric quantities in contracts.
1866: The United States Congress made metric units legal for trade and commerce in the U.S.A.
1871: The use of metric measure was made compulsory in the newly formed German Empire.
1875: The first International Metric Convention was held, Britain was not a signatory.

1884: Britain joined the International Metric Convention.
1897: Britain legalized the use of metric quantities in trade and commerce.
1935: International Electrotechnical Commission adopted Professor Giorgi's Rationalized MKSA System.

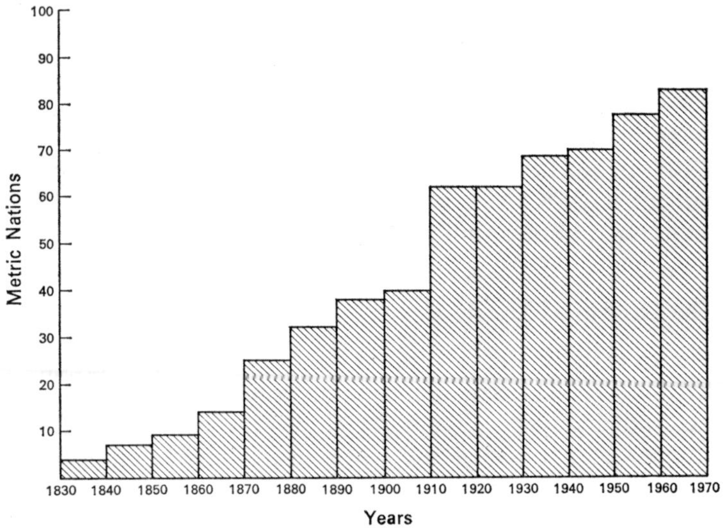

FIG. 2.1. Growth of world usage of the metric system.

1951: Hodgson Committee, sponsored by the Board of Trade, recommended that a change to metric units was desirable. They added that the change should be made in concert with other inch-using countries.
1954: General Conference on Weights and Measures added the degree Kelvin and the candela to the list of basic units.
1960: The MKSA System officially entitled System International (SI).
1965: Britain committed to adoption of SI units by 1975.

The sheer weight of evidence to support the case for a change to SI units is undeniable.

The United Nations lists 122 members: eighty-three of these are using the metric system. (China is a metric nation.)

Sixty-five per cent of the nations of the world are using the metric system.

India, Japan and China have changed since the Hodgson Committee reported.

Eighty-five per cent of the world's population live in metric nations.

Britain, South Africa, Australia and New Zealand are planning their changeover.

Seventy-five per cent of world trade is conducted in metric quantities.

Fifty-five per cent of Britain's exports go to metric countries.

Sixty nations are members of ISO and twenty-three nations make SI units the legal basis of their measures.

TABLE 2.1. AN ANALYSIS OF WORLD METRIC USAGE

(*Note:* some nations shown as Imperial users are in the process of changing to metric)

Name	System	Date of adoption	Name	System	Date of adoption
Aden	Imp		Bulgaria	M	1802
Afghanistan	M	1920	Burma	M	1920
Albania	M	1950	Burundi	M	
Algeria	M	1943			
Andorra	M		Cambodia	M	
Angola	M		Cameroon Rep.	M	
Argentina	M	1887	Canary Is.	M	
Australia	Imp/M		Cape Verde Is.	M	
Austria	M	1867	Central African Rep.	M	
Bahrain	Imp		Ceylon	Imp	
Belgium	M	1820	Chad Rep.	M	
Bolivia	M	1870	Chile	M	1865
Brazil	M	1862	China	M	1929
Britain	Imp/M	1970	Colombia	M	1854
Brit. Honduras	Imp		Congo Rep.	M	1910
Brunei	Imp		Costa Rica	M	1912

(*cont.*)

TABLE 2.1 (*cont.*)

Name	System	Date of adoption	Name	System	Date of adoption
Cuba	M	1858	Iran	M	1933
Cyprus	M	1961	Iraq	M	1933
Czechoslovakia	M	1876	Israel (Palestine)	M	1928
			Italy	M	1861
Dahomey	M		Ivory Coast	M	
Denmark	M	1910			
Dominican Rep.	M	1913	Japan	M	1909
			Jordan	M	1953
Ecuador	M	1856			
Egypt	M	1939	Korea	M	
Eire	Imp/M				
El Salvador	M	1886	Laos	M	1910
Eritrea	M		Lebanon	M	1935
Ethiopia	M		Libya	M	1920
			Liechtenstein	M	1870
Faroe Is.	M		Luxembourg	M	
Finland	M	1892			
Formosa	M		Macau	M	
France	M	1794	Malagasy Rep.	M	
French Guiana	M		Malaya (Singapore)	Imp/M	
French Polynesia	M		Malta and Gozo	M	1921
French Somaliland	M		Mali	M	
			Martinique	M	
Gaboon	M		Mauritania	M	
Germany	M	1872	Mauritius	M	1878
Greece	M	1922	Mexico	M	1928
Greenland	M		Monaco	M	
Guadeloupe	M	1912	Mongolia	M	
Guatemala	M		Morocco	M	1920
Guinea Rep.	M		Mozambique	M	
Haiti	M	1921	Muscat and Oman	Own system	
Honduras	M	1912			
Hong Kong	Imp				
Hungary	M	1876	Nepal	M	
			Netherlands	M	1820
Iceland	M	1907	Netherlands Antilles	M	
India (1966)	M	1920	New Caledonia	M	
Indonesia	M	1930			

TABLE 2.1 (*cont.*)

Name	System	Date of adoption
New Zealand	Imp/M	
Nicaragua	M	1912
Niger	M	
North Borneo	Imp	
Norway	M	1882
Pakistan	Imp/M	
Panama	M	1910
Paraguay	M	1901
Peru	M	1869
Philippines	M	1860
Poland	M	1910
Portugal	M	1872
Portuguese Guinea	M	
Puerto Rico	M	1840
Réunion	M	
Rumania	M	1884
Russia	M	1927
Rwanda	M	
Saint Marin	M	1900
Sao Tomé	M	
Sarawak	Imp	
Saudi Arabia	M	
Senegal	M	
Seychelles	M	1880
Somalia	M	
South Korea	Own system/M	1940
Spain	M	1859
Spanish North Africa	M	
Sudan	Own system/M	1950
Surinam	M	1910
Swaziland	Imp	
Sweden	M	1878
Switzerland	M	1875
Syria	M	1955
Taiwan	M	
Tanganyika	Imp	
Tangier	M	
Thailand	M	1923
Togo	M	
Tunisia	M	1895
Turkey	M	1934
United Arab Rep.	M	
United States	Imp/M	
Union of South Africa	Imp/M	
Upper Volta	M	
Uruguay	M	1894
Venezuela	M	1914
Vietnam	M	
Yugoslavia	M	1883
Zanzibar and Pemba	Imp	

2.6. The United States of America

The situation in the United States is that a strong body of opinion in favour of a changeover to the metric system has begun to influence Congress. A poll, conducted in 1967, among 3,100 technologists and scientists in American industry showed that 94% were in favour of a complete metrication policy, 3% were against and 3% wanted a partial change. It is interesting to note that the same poll revealed that 37% used only metric units already, 43% used both Imperial and metric and only 20% used only Imperial. These figures are quoted by permission of Industrial Research Incorporated, who conducted the survey and who are actively campaigning for American metrication. A Congressional Committee is considering the whole question and a number of shrewd guesses lead one to conclude that the United States may well start a metrication programme in 1972. A Metric Study Bill has been passed by the U.S. Congress and a similar Bill is before the Canadian Parliament.

CHAPTER 3

System International Units

3.1. Introduction to SI

The six arbitrarily defined basic SI units are shown in Table 3.1. From these six basic units and four supplementary units are derived all the units of the system.

The expression of quantities much larger or much smaller than the basic or derived unit is achieved in SI by compiling a list of multipliers, all of which are multiples or sub-multiples of ten. The multiples are given names to be used as prefixes to the name of the basic or derived unit: see Table 3.2. Positive powers of ten are given Greek names and negative powers take Latin names. This simple and consistent structure is one of the reasons for the rapid growth of world usage of the metric system in science and technology.

There is a flaw in the system; it is the fact that the unit of mass,

SI UNITS

TABLE 3.1. BASIC UNITS

Quantity	Name of unit	Unit symbol
Length	Metre	m
Mass	Kilogramme	kg
Time	Second	s
Electric current	Ampere	A
Thermodynamic temperature	Kelvin	K
Luminous intensity	Candela	cd

TABLE 3.2. PREFIXES AND MULTIPLIERS

Factor by which the unit is multiplied		Prefix	Symbol	Example
One million million (billion)	... 10^{12}	tera	T	teraohm (TΩ)
One thousand million	... 10^{9}	giga	G	gigahertz (GHz)
One million	... 10^{6}	mega	M	megawatt (MW)
1000	... 10^{3}	kilo	k	kilometre (km)
100	... 10^{2}	hecto	h	hectobar (hbar)
10	... 10^{1}	deca	da	decagramme (dag)
0.1	... 10^{-1}	deci	d	decimetre (dm)
0.01	... 10^{-2}	centi	c	centimetre (cm)
0.001	... 10^{-3}	milli	m	milligramme (mg)
0.000 001 (one-millionth)	... 10^{-6}	micro	μ	microsecond (μs)
One thousand-millionth	... 10^{-9}	nano	n	nanometre (nm)
One million-millionth (one-billionth)	... 10^{-12}	pico	p	picoampere (pA)
One thousand-million-millionth (one thousand-billionth)	... 10^{-15}	femto	f	femtofarad (fF)
One million-million-millionth (one-trillionth)	... 10^{-18}	atto	a	

Note. The words billion and trillion are used in their European sense, that is one billion is one million million, whereas in the United States one billion is one thousand million. One trillion means, in Europe, one million million million, whereas in the United States one trillion means one million million.

the kilogramme, has a name which includes a prefix. This fact creates an ambiguity in the use of multipliers with the gramme. A megagramme (Mg), for example, which is one million grammes, is one thousand times the basic unit of mass, and not one million times as the name might suggest. To overcome this objection it has been suggested that the kilogramme be renamed. The name Einstein, or Stein, has been suggested so that 1 kg = 10^3 g = 1 E. There is, at present, no official proposal to make the change.

In 1964 the old (1901) definition of the litre, as the volume of a kilogramme mass of water at 40°C, was abolished. The 1964 ruling is that the word 'litre' is now, by definition, merely a special name for a volume exactly equal to one cubic decimetre (dm^3). The old link-up between mass and volume is abolished.

For safety's sake, however, it is better to avoid the word 'litre' when referring to very precise measurements of volume and to make matters quite clear by referring only to cubic decimetres. For the same reason when ambiguity must be avoided one refers to cubic centimetres (cm^3) rather than to millilitres (ml). In the great majority of everyday commercial and technological cases the litre, with its multiples and submultiples, will be used.

3.2. Rules and Recommendations for Use of Prefixes

Rule (a) When a power index appears over a prefixed unit the index is taken to apply to the combined symbols thus:

$$1\ km^2 = 1\ (km)^2 = 10^6 m^2.$$

Rule (b) Prefixes should only be applied to the numerators of combined units which are of fractional form.
Thus MN/m^2 is preferred,
N/mm^2 is non-preferred.

Rule (c) The first choice of multiples and submultiples should be from those in which the power of ten is a multiple of three.
First choice to be $10^{\pm 3n}$, where n is an integer.

Rule (d) Double prefixes should be avoided, for example: Mm not k(km).

Some explanation and discussion of Rule (c) is necessary:

The metric system is provided with fourteen multipliers, a considerable burden on the memory. When the SI was adopted an attempt was made to rationalize the use of prefixes by the introduction of Rule (c). It is, however, a minor rule which is waived when there are good practical reasons for so doing. There are units in common use, for example the centimetre and decimetre, which are not SI preferred submultiples but which are unlikely to be suppressed. According to Rule (c) the time units of the minute and the hour are not SI but they will, of course, continue to be used.

3.3. The Basic Units

The three fundamental units of length, mass and time were introduced in Chapter 2.

The other basic units are those of temperature, electric current and luminous intensity.

The unit of thermodynamic temperature is the kelvin. The degree intervals of the Kelvin scale are identical with those on the Celsius scale, the bases of the two systems are, however, different. On the Kelvin scale the ice point of water is 273.15°K. On the Celsius scale the ice point of water is 0°C. The triple point of water is the fundamental fixed point for definition of the scale and is assigned the value 273.16°K exactly.

The candela is defined in terms of a full radiator ('black body') at the temperature of the solidification of platinum (2042°K). Under these conditions the radiator is said to have a luminous intensity of 60 candela per square centimetre.

The ampere is the constant current which, flowing in two long straight parallel conductors of small circular cross-section one metre apart *in vacuo*, causes them to exert a mutual force of 2×10^{-7} newton per metre length.

3.4. The Supplementary Units

To the six basic SI units are added supplementary units, named and defined as follows:

The radian is the unit for plane angular measure and is defined as the angle between two radii of a circle which cut off on the circumference an arc equal in length to the radius.

The steradian is the unit for solid angular measure and is defined as the solid angle having its vertex at the centre of a sphere and cutting off an area of the surface of the sphere equal to that of a square having sides of length equal to the radius of the sphere.

The curie is the unit for measurement of activity of radionuclides and is equal to an atomic disintegration rate of 3.7×10^{10} events per second. The coherent SI unit of activity is the reciprocal second, or one event per second, but the curie is a well-established unit and has been adopted as standard.

The mol, or mole, is the unit for the amount of substance in chemistry. One mol is defined as one gramme-molecule of the substance. In order to obtain a unit compatible with SI the kilogramme-mole may very well be adopted.

3.5. Coherence

The SI units belong to a coherent set. A coherent set of units is so arranged that the quotient or product of any two unit quantities leads to the unit of the resultant quantity. For example, when unit length (1 metre) is divided by unit time (1 second), unit velocity, 1 metre per second, results. One kilometre per hour, for instance, is not a member of the coherent set. The most important consequence of adopting a coherent set for SI is the introduction of the newton as the unit of force. The metric technical force was the kilogramme-force (kgf).

Let us consider the whole question of force, mass and weight: Newton's Second Law states that a particle, free to move, on which a force is acting experiences an acceleration in the direction of that force, proportional to the force and inversely proportional to the mass of the particle. Hence,

Force is proportional to mass times acceleration.

In the SI the unit of mass is 1 kilogramme,
the unit of length is 1 metre,
and the unit of time is 1 second.

Hence the unit of acceleration is

1 metre per second per second (1 m/s^2)

so that

1 coherent unit of force is that force which, acting on a mass of 1 kg, free to move, will give it an acceleration of 1 m/s^2.

This unit has been named the newton, symbol N. Thus:

$$1 \text{ newton} = 1 \text{ kg} \times 1 \text{ m/s}^2.$$

However, on earth a mass of 1 kilogramme will, if allowed to fall freely, accelerate at 9.806 65 m/s^2.

This acceleration due to gravity is denoted by 'g' and the value given is the average value obtained from measurements at many points on the Earth's surface. Local values may vary by up to 0.5%.

From the value of g we deduce that the Earth exerts a force equal to 9.806 65 newtons on a kilogramme mass on its surface.

We obtain, therefore, a relationship between the newton and the kilogramme force.

$$9.806\ 65 \text{ newtons} = 1 \text{ kilogramme force.}$$

At this point we must introduce the word 'weight'. Weight is the name given to the force due to the Earth's gravity acting on a mass on its surface.

The world of science and technology divides neatly into two parts: statics and dynamics. In the world of statics we are concerned with masses resting on supports on the surface of the Earth. The masses are considered to be at rest relative to the Earth and to each other. The only forces acting are those due to gravity. We may therefore, quite legitimately, use the idea of a kilogramme weight.

When we move into the world of dynamics, however, forces may be acting in all directions and due to a variety of causes. In this world coherence is a great simplifying aid and the use of the newton as the unit of force removes a multitude of troublesome 'g' factors from the equations. Pressure and stress, moment and torque, power and energy, are all derived quantities in which a coherent force unit is an advantage. It is in this world that the newton has a place.

3.6. Some Derived Units

A brief discussion of the units of SI for work, torque and power will serve to demonstrate the simplicity and freedom from coefficients which is characteristic of coherent systems:

Unit work in SI is 1 newton metre.

This unit has been named the joule. Thus:

$$1 \text{ Nm} = 1 \text{ joule (J)}$$

Unit power, which is rate of doing work, is

1 newton metre/second.

This unit has been named the watt (W):

$$1 \text{ Nm/s} = 1 \text{ J/s} = 1 \text{ watt (W)}.$$

Consider a torque of T Nm exerted through an angle θ radians.

The work done is $T\,\theta$ joules and the power is $T(\mathrm{d}\theta/\mathrm{d}t)$ watts, for $(\mathrm{d}\theta/\mathrm{d}t)$ we write ω, the angular velocity in radians per second, so that $T\omega$ = power in watts.

An engine or motor delivers power by exerting a torque along a rotating shaft. In a coherent system the unit of torque multiplied by the unit of angular velocity must give the unit of power. The cumbersome coefficients required for 'horsepower' formulae are unnecessary in SI.

TABLE 3.3. SOME DERIVED UNITS OF SI WITH COMPLEX NAMES

Physical quantity	Units of SI	Unit symbol
Area	Square metre	m^2
Volume	Cubic metre	m^3
Density (mass density)	Kilogramme per cubic metre	kg/m^3
Velocity	Metre per second	m/s
Angular velocity	Radian per second	rad/s
Acceleration	Metre per second squared	m/s^2
Angular acceleration	Radian per second squared	rad/s^2
Pressure	Newton per square metre	N/m^2
Surface tension	Newton per metre	N/m
Dynamic viscosity	Newton second per metre squared	Ns/m^2
Kinematic viscosity, Diffusion coefficient	Metre squared per second	m^2/s
Thermal conductivity	Watt per metre kelvin	W/(m K)
Electric field strength	Volt per metre	V/m
Magnetic field strength	Ampere per metre	A/m
Luminance	Candela per square metre	cd/m^2

Here it should be noted that it is neither possible nor desirable to distinguish between torque and work by reversing the symbols. One cannot write mN for metre newton as it could be misread as milli-newton.

Rotational speed is dealt with very simply in SI by thinking in terms of revolutions per second, for which the unit is the reciprocal second (s^{-1}).

The link-up with angular velocity is then simply

$$\omega = 2\pi\, n,$$

where n is in revolutions per second.

Vibrations are thought of as having a frequency in cycles per second, for which the unit is the hertz (Hz):

$$1 \text{ Hz} = 1 \text{ cycle/second } (s^{-1}).$$

The connection between vibration frequency and rotational speed is seen much more easily when rotational speed is quoted in revolutions per second and not in revolutions per minute.

3.7. The Hectobar

The SI unit for pressure or stress, that is force per unit area, is the newton per square metre (N/m^2).

In many engineering contexts the newton is rather small, being equal to 1/9.806 65 of a kilogramme force. This gives us an idea of the newton as being of the order of 100 grammes weight, the 'weight of a large apple'. The metre, on the other hand, is rather large; a square metre being the area of a normal domestic dining table. One newton per square metre is, therefore, a very small pressure. Normal atmospheric pressure, for instance, is about a hundred thousand newtons per square metre:

$$(101{,}325 \text{ N/m}^2).$$

The usual magnitude of stress met with in structural steels will be several tens of millions of newtons per square metre.

According to the rule for first choice multiples and submultiples we should use only kN/m^2, MN/m^2 and so on, using prefixes representing $10^{\pm 3n}$. However, this is a case where it appears that the rule is very likely to be waived, and preference given to a special multiple (10^7 N/m^2) called the hectobar (hbar).

The unit originates from the older cgs system thus:

In the cgs system 10^6 dyn/cm^2 is called a bar (short for barometric pressure).

The reason for the choice is that a 'standard atmosphere' is

$$1{,}013{,}250 \text{ dyn/cm}^2$$

so that 10^6 dyn/cm^2 is a convenient absolute unit in the cgs system, making the standard atmosphere 1.013 250 bar.

In SI terminology

$$1 \text{ bar} = 100 \text{ kN/m}^2$$

and

$$1 \text{ hectobar} = 10 \text{ MN/m}^2 = 10^7 \text{ N/m}^2.$$

Two arguments in favour of the hectobar are that it bears very simple relationships to the metric technical stress unit and to the imperial stress unit.

$$1 \text{ hbar} = 1.02 \text{ kgf/mm}^2$$
$$\doteqdot 1 \text{ kgf/mm}^2 \text{ within } 2\% \text{ error}$$

and

$$1 \text{ tonf/in}^2 = 1.544\ 43 \text{ hbar}$$
$$\doteqdot 1.5 \text{ hbar within } 3\% \text{ error.}$$

For most practical purposes therefore the hectobar and the kgf/mm^2 are interchangeable and to convert a tonf/in^2 stress all one need do is multiply mentally by $1\frac{1}{2}$ to obtain the approximate value in hectobar. Thus, a '40 ton' steel becomes a '60 hbar' steel. BSI advise that wherever an impure unit such as the hectobar is used, the pure unit of SI, in this case the MN/m^2, should be stated. Thus, 60 hbar (600 MN/m^2).

3.8. Fracture Toughness Testing

The adoption of the hectobar would determine the units to be used for quantities in which stress plays a part.

An important example is fracture toughness testing. The basic parameters for a quantitative discussion of fracture toughness are:

G = the strain energy release rate with crack extension per unit length of crack border, or crack extension force. This has the dimensions

$$\frac{L \times F}{L^2} = \frac{L \times MLT^{-2}}{L^2}.$$

In SI G could be expressed in

kJ/m² . . . kilojoules per square metre.

The quantities expressed in the literature as G_c and G_{1c} are critical values of G and will therefore also be in kJ/m².

K = the stress intensity factor in the vicinity of the crack front. This has the dimensions

$$\frac{F}{L^2} \sqrt{L}$$

which is

$$\frac{MLT^{-2} L^{\frac{1}{2}}}{L^2}.$$

In SI the accepted unit will be the

hectobar $\sqrt{\text{cm}}$
read as 'hectobar root centimetre'
or the bar $\sqrt{\text{cm}}$ read as 'bar root centimetre';

this has the advantage of emphasizing that K and its critical values K_c and K_{1c} are products of

stress and (length)$^{\frac{1}{2}}$.

If Young's Modulus is represented by E, then it is necessary that the equation

$$K = \sqrt{EG}$$

be satisfied and so the choice of units has to be made accordingly.

The units stated are consistent if E is in bars, or its multiples. The most convenient multiple of the bar to use for Young's Modulus is the megabar (Mb)

$$1 \text{ Mb} = 100 \text{ GN/m}^2.$$

An E of 13,000 tonf/in^2 converts to 2 Mb, approximately.

Some Continental writers have expressed the hectobar as the 'decanewton per square millimetre' (daN/mm^2). The symbol hbar is to be preferred both on grounds of economy and avoidance of prefixes in the denominator.

3.9. Units of Viscosity

There are two reasons for choosing viscosity units as an example for detailed discussion. The first is that there are numerous commercial viscometers which indicate kinematic viscosity in terms of 'seconds'. This value is the time for a known quantity of fluid to flow through a small orifice. This has led to a certain obscuring of the true dimensions of kinematic viscosity. The second reason is that viscosity units provide an excellent example of the conversion of derived units by dimensional transfer.

Dimensional analysis is the method by which we deduce information about a phenomenon from the single premise that the phenomenon can be described by a dimensionally correct equation connecting certain variables. The phrase 'dimensionally correct' implies that if the fundamental concepts of length, mass, time, force and so on are represented by symbols such as L, M, T, F, etc., then equations representing inter-relationships between groups of the symbols must balance, or be consistent.

Viscosity is the term applied to the internal friction of a fluid. It is the property which makes the application of force necessary to shear a fluid film. The first and classic description in quantitative terms was given by Isaac Newton. His experimental model consisted of two concentric cylinders submerged in a fluid. The radial difference between the cylinders was small so that the fluid in the annular gap formed a thin film, thin that is in comparison with its area. The force necessary to rotate one cylinder relative to the other gives an indication of the internal friction of the fluid. Newton found that the force was proportional to the swept area of the cylinder, proportional to the speed of rotation and inversely proportional to the film thickness.

Let the velocities of the cylinders be v_1 and v_2, so that their relative velocity is

$$v = (v_1 - v_2)$$

then

$$F = \mu A \frac{v}{h}$$

where F is the force necessary to maintain the relative velocity v, h is the film thickness and A the area of the film. v/h is the velocity gradient in the film and μ is a constant of proportionality. Newton found that μ was a true constant for a given fluid at one temperature. It is known as the coefficient of viscosity or the absolute viscosity. Usually v/h is a continuous function and may be replaced by $\mathrm{d}v/\mathrm{d}h$. Thus

$$F = \mu A \frac{\mathrm{d}v}{\mathrm{d}h}$$

so that

$$\mu = \frac{F}{A} \cdot \frac{1}{\mathrm{d}v/\mathrm{d}h}.$$

Hence the coefficient of absolute, or dynamic, viscosity is

$$\mu = \frac{\tau}{\mathrm{d}v/\mathrm{d}h}$$

where τ is the shearing stress and $\mathrm{d}v/\mathrm{d}h$ is the velocity gradient in a film of thickness h.

The dimensions of μ are

$$\frac{F \times T}{L^2},$$

force F having the absolute dimensions MLT^{-2}.

In SI μ has the units

$$\frac{\text{newton second}}{\text{metre}^2}.$$

Now, in the cgs system the unit was

$$\frac{\text{dyne second}}{\text{centimetre}^2}.$$

This unit was called the poise, but:

$$10^5 \text{ dyne} = 1 \text{ newton}$$
$$10^4 \text{ centimetre}^2 = 1 \text{ metre}^2$$

$$1 \text{ poise} = \frac{\text{dyn sec}}{\text{cm}^2} = \frac{10^{-5}\text{N.sec}}{10^{-4}\text{ m}^2}$$
$$= \frac{10^{-1}\text{ ns}}{\text{m}^2}.$$

In the Imperial pound-force-inch-second system μ is in

$$\frac{\text{lbf.sec}}{\text{in}^2}.$$

This unit was called the reyn. The practical unit was the microreyn.

It should be noted that the above reasoning applies only to Newtonian fluids. These are fluids which show a direct proportionality between stress and shearing velocity. In these fluids the shearing force may be reduced to very low values but there will still be a shear velocity.

There are materials, such as greases, plastic solids, ordinary oils below their pour-point temperatures, and some synthetic oils, which are not Newtonian. In these, some initial shear stress must exist before flow begins. The minimum stress to start flow is sometimes referred to as the yield value.

The property of more practical interest is the kinematic viscosity. Kinematic viscosity is the ratio of the absolute viscosity to the density:

$$\gamma = \mu/\rho.$$

The dimensions of kinematic viscosity are

$L^2\,T^{-1}$ and the cgs unit is the Stokes,

$$1\ \text{St} = 1\,\frac{\text{cm}^2}{\text{s}} = 100\ \text{centistokes.}$$

(Fluidity Φ is the reciprocal of the dynamic viscosity.)

In SI the unit of kinematic viscosity is metre²/second:

$$1\ \text{St} = 10^{-4}\ \text{m}^2/\text{s}$$

or

$$1\ \text{cSt} = 10^{-6}\ \text{m}^2/\text{s}.$$

The units of viscosity, and their conversion factors obtained by dimensional transfer are summarized in Tables 3.4 and 3.5.

Certain industrial capillary viscometers enable the kinematic viscosities of fluids to be estimated by noting the time in seconds for a given quantity of fluid to pass through the capillary tube.

The Saybolt Universal Viscometer, for example, is used in the U.S.A. The well-established SAE number system for automotive lubricant classification is based on Saybolt Universal Seconds. The relationship between Saybolt Universal Seconds and the SI units is given by the nomogram of Fig. 3.1.

Other instruments using the capillary flow principle are the Saybolt Furol Viscometer, also used in the U.S.A.; the Engler Viscometer as used in Europe; and the Redwood Nos. 1 and 2 as used in Britain.

Conversion tables, nomograms and graphs are published by the instrument manufacturers and appear in several engineering handbooks. An example is given in Fig. 3.1, a viscosity conversion chart reproduced from *Theory and Practice of Lubrication for Engineers* by Dudley D. Fuller (John Wiley).

3.10. The SI Electrical Units

The great founders of electromagnetic theory, Gauss and Maxwell, had a mathematical concept of electromagnetic phenomena. In order to make the phenomena at least apparently comprehen-

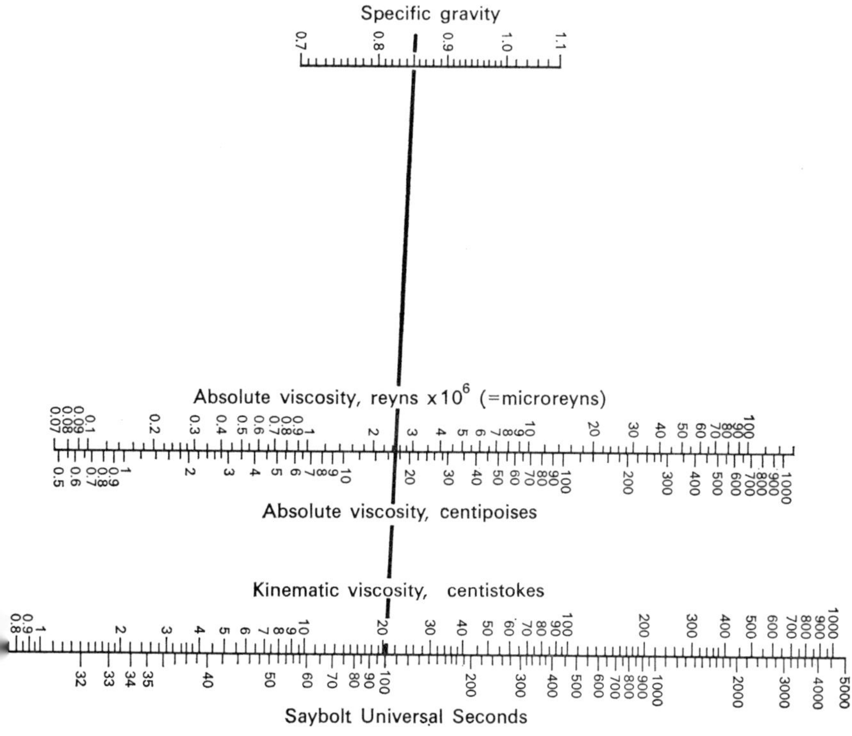

FIG. 3.1. Viscosity conversion chart. (Reproduced from Fuller, D. D., *Theory and Practice of Lubrication for Engineers*, John Wiley, New York, 1956.)

sible to lesser mortals, mechanistic notions were introduced and these, in conjunction with the application of the centimetre-gramme-second system of units, tended to give a false impression that electromagnetic phenomena were to be explained in mechanical terms. The simultaneous manipulation of three sets of units; cgs electromagnetic, cgs electrostatic, and 'practical', tended to obscure the underlying physics and created many stumbling blocks.

TABLE 3.4 UNITS OF ABSOLUTE OR DYNAMIC VISCOSITY

Dimensions	System International units	Centimetre Gramme Second units	Imperial Engineering units	Remarks
$MT^{-1} L^{-1}$	$\frac{\text{kilogramme}}{\text{metre second}}$	$\frac{\text{gramme}}{\text{cm second}}$	$\frac{\text{slug}}{\text{foot second}}$	In terms of mass
FTL^{-2}	$\frac{\text{newton second}}{\text{metre}^2}$	$\frac{\text{dyn second}}{\text{cm}^2}$	$\frac{\text{lbf second}}{\text{ft}^2}$	In terms of force
		$1 \frac{\text{dyn second}}{\text{cm}^2}$ = 1 poise	$1 \frac{\text{lbf second}}{\text{in}^2} = 144 \frac{\text{lbf second}}{\text{ft}^2}$ = 1 reyn	unit names
	$10^{-1} \frac{Ns}{m^2}$ =	1 poise =	1.45×10^{-5} reyn	Conversion factors 1 in² = 6.45 cm² 1 dyn = 0.225×10^{-5} lbf

TABLE 3.5 UNITS OF KINEMATIC VISCOSITY

Dimensions	System International units	Centimetre Gramme Second units	Imperial Engineering units	Remarks
$L^2 T^{-1}$	$\frac{\text{metre}^2}{\text{second}}$	$\frac{\text{cm}^2}{\text{second}}$	$\frac{\text{ft}^2}{\text{second}}$	
		$1 \frac{\text{cm}^2}{\text{s}}$ = 1 stokes		Unit name
	$10^{-4} \frac{m^2}{s}$ =	1 St =	$1.08 \times 10^{-3} \frac{ft^2}{s}$	Conversion factor 1 ft² = 929 cm²

It was Oliver Heaviside who first put forward the idea of 'rationalizing' electrical formulae in 1882. He had the support of J. A. Fleming and S. P. Thomson but no real notice was taken of his ideas for twenty years. In 1904 Professor Giorgi, of Rome, reintroduced the idea in conjunction with the MKS system.

The idea is to remove the factor 4π from formulae involving rectilinear systems and to allow it to appear in expressions relating to spheres, where its appearance is more logical. To do this one has to assign a value of 4π to μ_0, the magnetic constant of free space. In order to remove sundry powers of ten from electrodynamic formulae a factor of 10^{-7} is also included in the value of μ_0:

$$\mu_0 = 4\pi \times 10^{-7} \text{ henry/metre.}$$

Since it is an experimental fact that the electric space constant ϵ_0, also known as the permittivity of free space, is related to μ_0 by

$$\mu_0 \cdot \epsilon_0 = \frac{1}{c^2}$$

where c is velocity of light (3×10^8 m/s), ϵ_0 is also fixed and is equal to

$$\frac{1}{4\pi \times 10^{-7} \times (3 \times 10^8)^2} = 8.854\,16 \times 10^{-12} \text{ farad metre.}$$

The arrangement of the factors 4π and various powers of ten is then more convenient, enabling the basic laws to be simply stated.

There is nothing absolute about this. The Giorgi MKSA Rationalized System is a tidy one, which avoids obscuring the facts with cumbersome formulae. It is the homogeneity and coherence of the equations that is important.

The idea of a 'Characteristic Impedance of Free Space' arises because $\sqrt{(\mu_0/\epsilon_0)}$ has the dimensions of an impedance. The symbol for this quantity is Z_0 and its value in SI is 376.731 ohms.

Table 3.6 illustrates the appearance of some MKSA quantities and relationships in their rationalized and unrationalized forms.

TABLE 3.6. ELECTRICAL UNITS

Quantity or relationship	Rationalized form	Unrationalized form
Magnetic space constant μ_0	$4\pi \times 10^{-7}$	10^{-7}
Magnetic field strength at radius r external to a long straight conductor of circular cross-section	$\frac{I}{2\pi r}$	$\frac{2I}{r}$
Magnetic field strength on the axis, at the centre of a solenoid of length l and radius r	$\frac{NI}{(4r^2 + l^2)^{\frac{1}{2}}}$	$\frac{4\pi NI}{(4r^2 + l^2)^{\frac{1}{2}}}$
Electric space constant ϵ_0	8.854×10^{-12}	1.113×10^{-10}
Capacitance of two parallel plates separation l and cross-section A	$\frac{A\epsilon}{l}$	$\frac{A\epsilon}{4\pi l}$

3.11. Specific Heat

The science of heat has suffered almost as much as has the science of electromagnetics from a profusion, and indeed a confusion, of units and unit systems.

In his book *Theory of Heat* James Clerk Maxwell wrote that special units of heat, which are really special units of energy, are an encumbrance to the science of heat. No other science employs a special unit of energy.

We should consider heat to be expressed in units of mechanical work (joules in System International). The calorie certainly should be abandoned.

If we set out the reasoning leading to the unit definitions in two columns, one in dimensional form, and one in unit form, we arrive at Table 3.7.

Now, the amount of heat Q required to raise the temperature of a body by θ is proportional to θ if we assume that the physical properties of the body remain constant,

$$Q \propto \theta.$$

TABLE 3.7. UNIT DEFINITIONS.

Quantity	Dimension symbol	System International	
		Unit symbol	Unit name
Length	L	m	metre
Mass	M	kg	kilogramme
Time	T	s	second
Temperature	θ	K	kelvin
Force	$F = M \times L \times T^{-2}$	N	newton
Energy	$E = F \times L$ $= M \times L^2 \times T^{-2}$	J	joule

(The interval on the kelvin scale is identical with the interval °C.)

If a constant, C, is introduced,

$$Q = C \,.\, \theta.$$

C may be termed the THERMAL CAPACITY of the body. The thermal capcity of the body must be proportional to the mass of the body;

$$C \propto m \quad \text{or} \quad C = c \,.\, m,$$

where c is the specific heat capacity of the body. Thus

$$Q = c \,.\, m \,.\, \theta.$$

We have already decided to express Q in joules when working in SI units and so we shall have:

$$J = c \,.\, \text{kg°C}.$$

Hence:

$$c = \text{J/kg°C}.$$

Now, the cgs system calorie is a water-based, artificial heat (energy) unit, and therefore comes under Maxwell's heading of 'encumbrance'.

The calorie is defined as the heat required to raise 1 g of water by 1°C at 15°C. This definition would give

$$1 \text{ calorie} = 4.185\,5 \text{ joules}.$$

The rounded value,

$$1 \text{ cal}_{IT} = 4.186 \text{ joules},$$

is derived from a mean value from international steam tables and is known as the international table calorie, or simply 'table calorie'.

This conversion factor, 'Joules Mechanical Equivalent of Heat', is the link between the artificial heat units and the 'real' energy units of dynamics.

The other commonly used unit systems define their artificial heat unit in the same way:

The Imperial Engineering System, for example, defines the British Thermal Unit as the heat required to raise 1 lb (mass) of water by 1°F. The official definition stipulates a mean value, 1/180th part of the heat required for a change from 32°F to 212°F.

Joules equivalent in the Imperial Engineering System is:

$$1 \text{ B.t.u.} = 778 \text{ ft lbf.}$$

When these water-based artificial heat units are used the specific heat of a substance is defined as the ratio of the heat required to raise unit mass of the substance through unit temperature interval to the heat required to raise unit mass of water through unit temperature interval. Given this definition, in the form of a ratio, specific heat is a dimensionless number and will be the same for any system of units which defines unit heat in relation to water.

SI, however, does not use a special heat unit, and in order to express specific heat in terms of the number of joules required to raise 1 kg of the substance through 1°C we must multiply the specific heat in cal/g°C by 4.186×10^3.

There still remains the possibility of retaining the ratio definition, in which case the heat required for the substance and the heat required for water would both be multiplied by Joules Equivalent and the ratio would, of course, still be the same dimensionless number.

The J kg^{-1} deg^{-1}C form is the correct SI one.

CHAPTER 4

National and International Standards

4.1. National Standards

During the early years of the twentieth century each major industrial nation recognized the need for a national body to be responsible for the compilation of standards and codes of practice. The purpose of the standards and codes of practice was to prevent the proliferation of varieties and sizes of the products of industry and to avoid the inefficiency associated with excessive variety. An important consequence of the establishment of standards is that the interests of the user as well as the manufacturer have to be considered and so questions of quality, methods of test and definitions of adequate performance are brought within the scope of the standards.

The national standards organizations are listed as follows:

(*Note.* The symbol M implies that metric standards are issued. This is sometimes the case even where the metric system is not the primary system of measure.)

(The initials in brackets are the commonly accepted designations for the organizations listed.)

UNITED KINGDOM:

British Standards Institution (BSI)

AUSTRALIA:

Standards Association of Australia (AS)

CANADA:

Canadian Standards Association (CSA)

Canadian Govt. Specifications Board (CGSB)

CENTRAL AFRICA:

Standards Association of Central Africa (CAS) M

SOUTH AFRICA:

South African Bureau of Standards (SABS)

UNITED STATES OF AMERICA:

United States of America Standards Institute (USASI)
Federal Specifications and Federal Standards
Underwriters Laboratories Inc. (UL)
U.S. Dept. of Commerce
National Bureau of Standards (NBS)

BELGIUM:

Institut Belge de Normalization (NBN) M

BULGARIA:

Institut de Normalization, Mesures et Appareils de Mesure (BDS) M

CUBA:

Dirección de Normas y Metrologia (UNC) M

FRANCE:

Association Française de Normalisation (AFNOR) M
Union Technique de l'Électricité (UTE)

GERMANY:

Deutscher Normenausschuss (DIN) M

ISRAEL:

Standards Institution of Israel (SI) M

ITALY:

Ente Nazionale Italiano di Unificazione (UNI) M
Comitato Elettrotecnico Italiano (CEI)

NORWAY:

Norges Standardizerings—Forbund (BS) M

POLAND:

Polski Komitet Normanlizacyjny (PN) M

PORTUGAL:

Reporticao de Normalizacas (NP) M

RUSSIA:

Gosudarstvennyi Komitet Standartov (GOST) M

SWEDEN:

Sveriges Standardizerings Kommission (SIS) M

4.2. International Standards

The first attempts at rationalization were made between small groups of neighbour nations or between small groups of nations with some other strong common interest. Examples of this process are:

(a) Euronorm: this organization is the standardizing body of the European Coal and Steel Community, formed by Belgium, France, Germany, Holland, Italy and Luxembourg.

(b) The ABC countries: this organization, formed by America, Britain and Canada, evolved the unified screw thread system and was also largely responsible for establishing the Imperial version of the ISO system of limits and fits as the common standard between the nations named.

The first major international effort towards establishing a common language of measurement was the International Metric Convention, formed in 1875. Britain joined the Convention in 1884, having legalized the use of metric quantities in legal contracts in 1864. America made metric quantities legally acceptable in 1866. These early efforts were concerned only with the units of measurement and not directly with standardization of products.

Real progress towards international standardization began in 1926 when the International Federation of the National Standardizing Associations (ISA) was formed by about twenty nations. The work of ISA was disrupted in 1939 after valuable work had been done, notably the ISA system of limits and fits. In 1944 the United Nations Standards Co-ordinating Committee (UNSCC), formed by eighteen nations, succeeded the ISA. In 1946 the UNSCC formed the International Standards Organization (ISO).

4.3. The International Standards Organization

The seat of the ISO is in Geneva, Switzerland. The present membership is as tabulated on p. 38.

Albania
Argentina
Australia
Austria
Belgium
Brazil
Bulgaria
Canada
Ceylon
Chile
Colombia
Cuba
Czechoslovakia
Denmark
Finland
France
Germany
Ghana
Greece
Hungary
Icaiti (C. America)
India
Indonesia
Iran
Iraq
Ireland
Israel
Italy
Japan
Korea N.
Korea S.
Lebanon
Mexico
Morocco
Netherlands
New Zealand
Norway
Pakistan
Paraguay
Peru
Poland
Portugal
Rumania
Singapore
South Africa
Spain
Sweden
Switzerland
Thailand
Turkey
United Arab Rep.
United Kingdom
U.S.A.
U.S.S.R.
Venezuela
Yugoslavia

Correspondent Members

Cyprus
Kuwait
Madagascar
Malta
Syria
Some sixty nations

The fact that powerful inch-using nations are ISO members means that, to some extent, the establishment of standards is a two-way traffic. Important American-based industries, such as oil and electronics, tend to impose their inch-based standards on the metric world. Many pipe and pipe-thread standards are inch-based. Standards for electronic printed circuit boards and electronic components based on American practice are widely used in metric countries.

We are thus led to the important conclusion that the primary object of this entire exercise is to 'go international'. This involves 'going metric' because eighty-three of the world's 122 nations use the metric system. Where well-established and accepted international standards exist, based on Imperial units, there is no really convincing reason for abandoning them. Such standards as the

British Post Office 19-inch rack for electronic equipment, printed circuit boards with contact strips at 0.1-inch pitch, inch-based pipe sizes and vehicle wheel-size standards need not be abandoned. They need only be made 'bilingual' by placing the metric equivalent in parenthesis or in italics after the Imperial dimension. In the long term, since metric units will be the primary ones, truly metric standards will gradually find acceptance but this will not be a rapid process.

4.4. Creation of the ISO

Discussions took place in London on 14th October 1946 between sixty-four delegates from twenty-five countries. The outcome of the UNSCC meeting was a decision to hold a provisional General Assembly of the ISO on 24th October, also in London. At the first General Assembly the ISO Constitution and Rules of Procedure were adopted. By 23rd February 1947 fifteen national standardization committees had ratified the provisions of the first General Assembly and the ISO was able to start work. Other national standardizing committees have since ratified the Constitution and Rules of Procedure, and other standardizing organizations have been admitted to membership.

4.5. Membership of the ISO

ISO Members are the national bodies most representative of standardization, one for each country. They are admitted on condition that they agree to abide by the ISO Rules of Procedure.

Organizations in developing countries which do not have a standards body may become Correspondent Members. The status of the applicant organization is investigated by the ISO before Correspondent Membership is granted.

4.6. Finance of the ISO

The ISO is financed by contributions from Member nations, the amount payable being calculated according to the importance of the country concerned.

Correspondent Members pay a symbolic annual subscription.

4.7. Objectives of the ISO

The primary objective of the ISO is to facilitate the international exchange of goods and services. As a means to this end the Organization takes action to co-ordinate and unify national standards and it issues recommendations to Member Bodies. An international standard may be set up providing no Member Body dissents. This procedure is not, however, invoked. The work of the Organization proceeds quite adequately by the issue of recommendations. The ISO is also at the disposal of other international organizations which may require specialized information on matters relating to technology and standardization.

4.8. The Structure of the ISO

The Organization consists of a General Assembly and Council, a President, Vice President and Treasurer. There also exists a Secretariat presided over by a General Secretary, located in Geneva, Switzerland. The detailed work of standardization is undertaken by Technical Committees and Technical Divisions. The General Assembly of Delegates nominated by Member Bodies meets at least once every three years.

The administrative organ of the ISO is the Council, consisting of fourteen representatives of Member Bodies. The council meets at least once every year to enquire into the activities of the Organization and to report to the General Assembly. The Council elects a Supervisory Committee to assist the President and to supervise the General Secretariat.

The Secretary-General keeps all Member Bodies, as well as the Council, informed of the work carried out by technical Committees.

4.9. The Technical Committees of the ISO

The Technical Committees are composed of a delegation from each of the Member Bodies wishing to take part in the work. The Secretariat to each committee is provided by a Member Body designated by the Council. The Member Body entrusted with a Secretariat has exactly the same status on the committee as the other members and must act impartially. ISO technical committees cover most fields of technology and science excluding electricity. All questions of an electrotechnical nature are dealt with by a separate and autonomous body, the International Electrotechnical Commission.

4.10. The International Electrotechnical Commission

The IEC is older than the ISO, having been formed in 1906. The members of the commission are the National Committees, one for each country, which represent the electrical interests of the country concerned: manufacturers, government authorities, teaching and professional bodies and users. Most national committees are recognized and supported by their respective national government.

The IEC issues recommendations and reports drawn up by its technical committees. The recommendations are intended to serve as the basis for national standards. The reports are published as an intermediate step in the standardizing process and serve as the basis for discussions, from which recommendations may be drawn up.

CHAPTER 5

Additional Benefits of Metrication

5.1. Access to International Standards

The simple fact of using the same basic units of length, mass and time as the majority of mankind will improve Britain's prospects of earning a living. There are, however, several important additional advantages to be gained by adopting ISO practice in its entirety. Access to a large body of established international standards will encourage manufacturers to rationalize their products and improve compatibility with existing practice.

5.2. Preferred Numbers

Practically all the products of industry have to be manufactured in a range of sizes, outputs or ratings. In 1877 Colonel Charles Renard pointed out that, in the vast majority of cases, experience showed that the range of values most likely to meet all needs was that in which a geometric progression was followed. A geometric progression is one in which each term in the series is obtained by increasing the preceding term by a constant percentage. This arrangement links the interval to the magnitude of the term, or in plain words the steps get progressively larger as the numbers get larger. Figure 5.1 shows the appearance of the internationally agreed series when they are plotted on a linear scale. The agreed series are known as Renard Series and are obtained by taking the following common ratios:

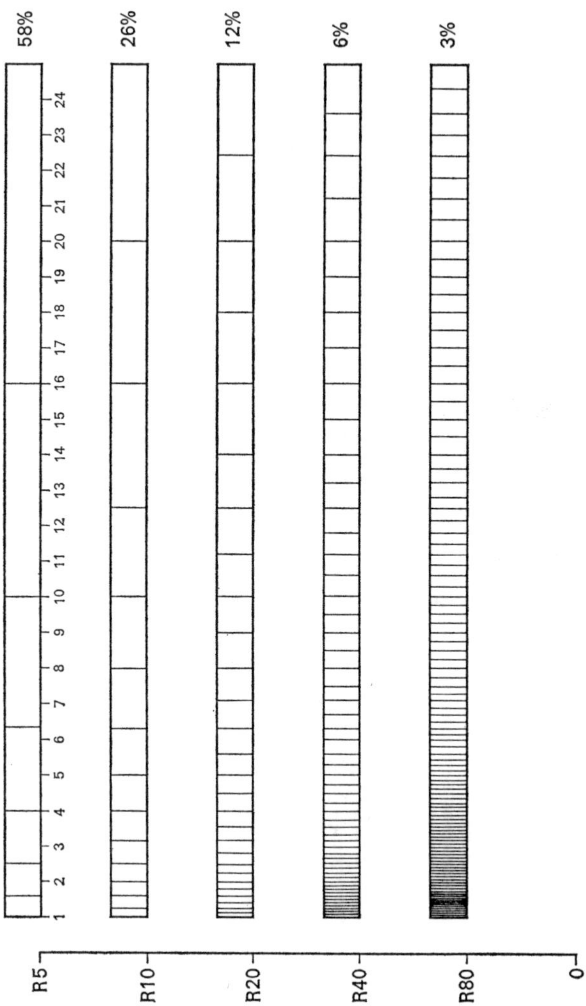

FIG. 5.1. The Renard system of preferred numbers.

Fifth root of ten	R5 Series
Tenth root of ten	R10 Series
Twentieth root of ten	R20 Series
Fortieth root of ten	R40 Series
Eightieth root of ten	R80 Series

This is equivalent to saying that the R5 series is obtained by dividing each decade into five steps geometrically, the R10 series by dividing each decade into ten steps geometrically and so on.

The common ratios as stated are equivalent to 1.58, 1.26, 1.12, 1.06 and 1.03. This means that successive terms increase by:

58% in the R5 Series
26% in the R10 Series
12% in the R20 Series
6% in the R40 Series
3% in the R80 Series

Table 5.1 shows the calculated values of the Renard Series. In view of the possible odd appearance of some of the numbers a rounding off is permitted. Table 5.2 shows rounded values which give terms that double every three terms in the R10 Series, every six terms in the R20 Series and every twelve terms in the R40 Series. The maximum permitted roundings are + 1.26% and − 1.01%. Any series can be extended upwards or downwards by multiplying or dividing by 10 any number of times. Additional series can be obtained by taking every second term of a series, or every third term, or every fourth term, etc. The resultant series are denoted by R5/2, R10/3, etc.

The designer planning a range of products has, therefore, only to decide how coarse or how fine should be the series he requires. Once that decision has been made and the largest and smallest values have been fixed the intermediate values are automatically settled. The application of the Renard Series is by no means universal. Continental and British metric practice will be to apply the rounded Renard numbers to products of circular cross section; bar, rod, wire and products made from them. Sheet and strip

TABLE 5.1. BASIC SERIES OF PREFERRED NUMBERS

1	2	3	4	5
Serial number	Basic series			
	R5	R10	R20	R40
0	**1.00**	**1.00**	1.00	1.00
1				1.06
2			1.12	1.12
3				1.18
4		**1.25**	1.25	1.25
5				1.32
6			1.40	1.40
7				1.50
8	**1.60**	**1.60**	1.60	1.60
9				1.70
10			1.80	1.80
11				1.90
12		**2.00**	2.00	2.00
13				2.12
14			2.24	2.24
15				2.36
16	**2.50**	**2.50**	2.50	2.50
17				2.65
18			2.80	2.80
19				3.00
20		**3.15**	3.15	3.15
21				3.35
22			3.55	3.55
23				3.75
24	**4.00**	**4.00**	4.00	4.00
25				4.25
26			4.50	4.50
27				4.75

(*cont.*)

TABLE 5.1 (*cont.*)

1	2	3	4	5
Serial number	Basic series			
	R5	R10	R20	R40
28		**5.00**	5.00	5.00
29				5.30
30			5.60	5.60
31				6.00
32	**6.30**	**6.30**	6.30	6.30
33				6.70
34			7.10	7.10
35				7.50
36		**8.00**	8.00	8.00
37				8.50
38			9.00	9.00
39				9.50
40	**10.00**	**10.00**	10.00	10.00

(Reproduced from BS Handbook 18: 1966, p. 50).

products, however, do not follow Renard Series. British practice will probably be to follow the established ranges of sheet and strip thickness regardless of whether or not they are Renard numbers.

Such criticism as has been levelled at the Renard Series tends to concentrate on the upper end of the range where, it is said, the steps are rather large. Proposals have been put forward for a modified series known as a subgeometric progression. In a subgeometric progression the magnitude of the steps increases monotonically while the ratio of terms decreases. The resulting series increases more rapidly than an arithmetic series but less rapidly than a geometric. There is, as yet, no official support for the idea.

TABLE 5.2.

Column	1		2			3			4		5
Number of terms or index	5		10			20			40		Serial number
Approximate ratio	1.6		1.25			1.12			1.06		
Series	R5	R″5	R10	R′10	R″10	R20	R′20	R″20	R40	R′40	
	1		1			1.0			1.0		0
									1.06	1.05	1
						1.12	1.1		1.12	1.1	2
									1.18	1.2	3
			1.25		(1.2)	1.25		(1.2)	1.25		4
									1.32	1.3	5
						1.4			1.4		6
									1.5		7
	1.6	(1.5)	1.6		(1.5)	1.6			1.6		8
									1.7		9
						1.8			1.8		10
									1.9		11
			2			2.0			2.0		12
									2.12	2.1	13
						2.24	2.2		2.24	2.2	14
									2.36	2.4	15
	2.5		2.5			2.5			2.5		16
									2.65	2.6	17
						2.8			2.8		18
									3.0		19
			3.15	3.2	(3)	3.15	3.2	(3.0)	3.15	3.2	20
									3.35	3.4	21
						3.55	3.6	(3.5)	3.55	3.6	22
									3.75	3.8	23
	4		4			4.0			4.0		24
									4.25	4.2	25
						4.5			4.5		26
									4.75	4.8	27
			5			5.0			5.0		28
									5.3		29
						5.6		(5.5)	5.6		30
									6.0		31
	6.3	(6)	6.3		(6)	6.3		(6.0)	6.3		32
									6.7		33
						7.1		(7.0)	7.1		34
									7.5		35
			8			8.0			8.0		36
									8.5		37
						9.0			9.0		38
									9.5		39
	10		10			10.0			10.0		40

Preferred numbers | More rounded values: 1st rounding | 2nd rounding |
Values in brackets to be avoided wherever possible

(Reproduced from BS Handbook 18: 1966, p. 55.).

5.3. ISO System of Limits and Fits

One of the basic problems of modern industry is that of interchangeability. Large numbers of components produced in many different establishments may be brought together and required to fit together without size adjustment. Components may have to be produced and dispatched to any part of the world as spare parts for assemblies made and sold long ago. The practical solution to the problem of interchangeability is to produce all the components in an assembly to a system of limits and fits. Many such systems have been devised but the most modern, comprehensive and universally accepted system is the ISO System described in ISO Recommendation 286. British Standard 1916: 1953 is based on the ISA System which was the forerunner of the ISO System. The differences between the ISA and ISO Systems are small and only relate to the extreme ends of the range of sizes covered. B.S. 1916: 1953 Part 2: *Guide to the Selection of Fits* is a useful explanatory booklet on the system in its inch and metric forms. The current standard used in the U.S.A. is ASA B41: 1955. It is equivalent to BS 1916. It is therefore true to say that the ISO System is the accepted one throughout the world. The technical details are discussed in Chapter 8.

5.4. ISO Metric Screw Threads

Fasteners are fundamental to all engineering and the haphazard growth of technology throughout the world has led to the production of some 210 different screw thread systems. This situation is so wasteful and chaotic that practically all nations are seizing the opportunity provided by the recently (1962) established ISO Metric range of threads. It is in the world of fasteners and screw threads that the greatest benefit of ISO standards will be felt both economically and in terms of convenience. Governments, in their capacity as large-scale purchasers of the products of industry, can bring great pressure to bear on industry by specifying ISO Metric fasteners in all new contracts. A recent poll of eighteen nations revealed that sixteen were adopting the policy of establishing ISO Metric threads as the primary system.

The technicalities of ISOM threads are discussed in Chapter 9.

CHAPTER 6

Precise Conversion of Imperial to Metric (SI) Quantities

6.1. Conversion of Simple Linear Dimensions

The word simple in this context implies an untoleranced dimension. There is, however, an implied tolerance in any numerical statement. The implied tolerance is always the unit of the last significant figure, unless otherwise stated.

For example:

78 inches

implies a tolerance zone 1 inch in extent symmetrically disposed about the nominal length; that is to say 78 inches plus or minus half an inch.

$$78.0'' \begin{array}{l} + 0.5'' \\ - 0.5'' \end{array}$$

If the figure 78 inches is to be read in any other way, then the desired tolerance must be stated alongside the figure or, as a general note, elsewhere on the document. One often finds such notes as:

'Tolerances on inch and fractional dimensions:

6 in. and below $\pm \frac{1}{64}$ in.
above 6 in. $\pm \frac{1}{32}$ in.'

In which case, of course, the 78-in. example is read as:

$78\frac{1}{32}$ in. maximum
$77\frac{31}{32}$ in. minimum.

6.2 Equivalent Nominal Designations

Equivalent nominal designations are those in which brevity is required, even at the expense of precision. For example, we know that 1 inch is equal to 25.4 millimetres exactly, this being the present definition of the British and American inch.

We would, however, be justified in referring to 1-inch bar stock as 25-millimetre bar stock. This is a convenient short-hand but there are potential traps for the user and it is wise to check the size of material described in such round figures to the degree of accuracy appropriate to the use to which the material will be put.

6.3. Dimensions for Functional Interchangeability

Where components are to be manufactured to either Imperial or metric dimensions, and are required to be interchangeable, there must obviously be very careful consideration of what is meant by a stated tolerance. The methods by which the component will be inspected and measured must also be considered.

In all practical cases the purely metrological problems are associated with the cost of production of the component concerned. One cannot tighten a tolerance without increasing the cost of production and the application of rigid rules derived solely from metrological considerations results, all too often, in unreasonable increases in production costs.

Having raised an array of problems, let us now consider their solution with the aid of examples.

First of all let us consider a toleranced linear dimension:

$$0.06 \text{ in.} \pm 0.005 \text{ in.}$$

The tolerance is 0.01 in. The dimension may be written:

0.065 in. maximum
0.055 in. minimum.

British Standard 2856: 1957, *Precise Conversion of Inch and Metric Sizes on Engineering Drawings*, provides us with a set of rules and a table for converting and rounding such dimensions. Table 6.1 gives fineness of rounding.

TABLE 6.1. FINENESS OF ROUNDING

1	2	3
Original tolerance in inches		Fineness of rounding. Derived millimetre sizes to be rounded to the nearest
From	Up to but not including	
in.	in.	mm
0.000 01	0.000 1	0.000 01
0.000 1	0.001	0.000 1
0.001	0.01	0.001
0.01	0.1	0.01
0.1	1.0	0.1

Rule 1. From the tolerance determine the fineness of rounding the converted value and accordingly the number of decimal places to be retained. The fineness of rounding is obtained from a table given in BS 2856 and reproduced in BS Handbook 18:1966, *Metric Standards for Engineering* (see Table 6.1).

In the example the fineness of rounding should be

0.01 mm.

Rule 2. Convert the limits of size in one system of units into the equivalent limits of size in the other system of units, using the conversion factor

1 in. = 25.4 mm exactly

and retaining temporarily two more decimal places than the rounded number of places determined by Rule 1.

In the example we have

0.065 in. max = 1.651 0 mm max.

Note: we have retained two figures after the rounding figure (0.01 mm).

0.055 in. min = 1.397 0 mm min.

Rule 3. Round the converted limits of size obtained by Rule 2, to the fineness determined by Rule 1, choosing the rounded value that is nearest the unrounded value.

If the unrounded value falls midway between two adjacent rounded values, choose the even rounded value.

In the example we have:

Maximum limit of size	1.651 0 mm
which rounds to	1.65 mm by Rule 1.
Minimum limit of size	1.397 0 mm
which rounds to	1.40 mm by Rule 1.

But, let us examine the implications in our converted values:

$$\text{Tolerance} = 1.65 - 1.40 = 0.25 \text{ mm.}$$

The procedure described suggests that 0.01 mm is really significant in the result, 0.01 mm being the unit of the last figure shown. We have declared the tolerance to be not 0.24 or 0.26, but 0.25 mm. Now, 0.01 mm is 0.000 4 in., very nearly. 0.000 4 in. is only 4% of 0.01 in., which was the original tolerance.

We have thus, by following the rules laid down in BS 2856, expressed the result in a way which implies that 4% of the original tolerance is significant. The designer is unlikely to have meant this. The procedure is, from a purely metrological point of view, a safe one. There are, however, other considerations, notably cost of production and of gauge making.

The crux of the matter is this:

1 mm is very nearly $\frac{1}{25}$, or 4% of 1 in.

It follows that, since,

	1 mm	represents a length equal to 4% of 1 in.,
then	0.1 mm	,, ,, ,, ,, ,, 4% of 0.1 in.
and	0.01 mm	,, ,, ,, ,, ,, 4% of 0.01 in.

and so on.

If a dimension in inches is expressed to a certain number of decimal places, and this dimension is converted into millimetres and the result stated to the same number of decimal places, then the final digit in the millimetre version represents a length only 4% of the length represented by the final digit in the inch version.

An attempt to work to such a dimension will entail increased costs in production and in gauge making. We must attempt to find a conversion procedure which will enable us to express the result in a way which does not imply greater precision than was originally called for.

6.4. An Alternative Conversion Procedure

Suppose we follow the rule that the dimension in millimetres shall always have one less place of decimals than the dimension in inches, the rule for correct rounding still being observed.

This procedure is still a safe one in most practical applications since:

	1 mm	represents a length equal to 4% of 1 in.
or	1 mm	,, ,, ,, ,, ,, 40% of 0.1 in.
	0.1 mm	,, ,, ,, ,, ,, 40% of 0.01 in.

so that if a dimension in inches is expressed to a certain number of decimal places, and this dimension is converted into millimetres and the result stated to one less place of decimals, then the final digit in the millimetre version represents a length only 40% of the length represented by the final digit in the inch version.

The example now becomes:

given dimension: 0.06 in.
tolerance: 0.01 in.

convert to metric and retain one decimal place:

maximum limit of size: 1.7 mm
minimum limit of size: 1.4 mm
tolerance: 0.3 mm.

The implications in this result are closer to the original statement than when the same number of decimal places appear in metric and inch versions. The discussion presented above is generally applicable. Conversion of pressures, areas, powers, and so on, can be examined in the same way and unjustifiable precision in the stated result can be avoided.

6.5. Other Considerations When Rounding

In some applications coarse rounding is justified because of the measuring device used or the nature of the unit.

To quote two examples:

A workshop metric micrometer, in good condition, might possibly be accurate to within plus or minus three micrometres (microns). To quote a figure implying that one micron was significant would not make sense to the workshop unless a special note on the measuring method to be used was issued.

In a different sense entirely it may be quite useless to give more than a set number of decimal places because of the discontinuous nature of the units involved:

Consider travelling expenses in pence per mile being converted into new (decimal) pence per kilometre.

$$1d. = 0.416\text{p}$$
$$\text{and } 1 \text{ mile} = 1.609\ 34 \text{ km}$$
$$\text{so that } 1d/\text{mile} = 0.416/1.609\ 34 \text{ p/km}$$
$$= 0.26\text{p/km}.$$

Consider the commonly occurring values:

$9d$/mile . . . this becomes (0.26×9) p/km
which is £0.023 4/km,
which has to be rounded *up* to $2\frac{1}{2}$p/km.

The cumbersome expression £0.023 4 for 2.34 new pennies is necessary because if one writes 2.34p this will be read as £2 and 34 new pennies.

The construction industry offers some examples of coarse rounding:

The fifth edition of the *Standard Method of Measurement of Building Works* appeared in metric form in mid-1968. The general Rules state that each measurement shall be taken to the nearest 10 mm. The rounding rule will be that over 5 mm will be rounded up to 10 mm and under 5 mm will be disregarded. It is therefore proposed to work with only two places of decimals of the metre.

6.6. Measuring and Gauging

The fact that the millimetre is approximately one twenty-fifth of an inch means, as we have already seen, that a quantity converted from Imperial to metric has an implied precision two and a half times that of the information given if we follow the 'one less decimal place' rule and twenty-five times that of the information given if we follow the rules of BS 2856.

Now: the tolerances allowed to measuring and gauging instrument device makers are in the region of 5% to 10% of the tolerances on the articles to be measured.

If designers are not made fully aware of the pitfalls in conversion they are likely to make unrealistic demands on the gauge maker.

CHAPTER 7

International Standard Drawing Office Practice

7.1. Projection Conventions

First Angle Projection is the primary system in Europe and the United Kingdom. Third Angle Projection is the primary system in the U.S.A. and Canada. The general desire in British industry is to use both systems without a firm commitment either way.

The ISO Recommendation No. 128: 1959, *Principles of Presentation*, is illustrated with First Angle Drawings, but a note admitting Third Angle Projection is included.

7.2. Drawing Sheet Sizes

The choice of drawing sheet sizes is somewhat complicated by the popularity of the 60 × 40-in. board, which takes an A0 metric drawing sheet, but does not take a 2A0 sheet.

The A0 size is the largest commonly used metric size and many users with 60 × 40-in. boards find it convenient to use metric A series sheets up to and including A0 and, for the largest sheet, a non-standard 60 × 40-in. sheet.

The 60 × 40-in. board is not restricted to the U.K.; European users find it convenient and refer to it as a 'Hamburg' size board.

Table 7.1 shows the range of drawing sheets available, and Table 7.2 shows the sizes of drawing boards in relation to drawing sheets in inch and metric sizes.

2A0 has been found to be inconveniently large and has the

disadvantages of not fitting the 60 × 40-in. or Hamburg boards. Practice seems to be evolving along the lines of keeping A sizes up to and including A0 and then keeping a stock of 60 × 40-in. sheets for drawings requiring the full use of the board.

A snag arises in storage. The usual plan chest is of Antiquarian size, 58 × 35-in. (1,473 × 889) overall, but having drawers of internal size 54½ × 32¼-in. (1,384 × 819). The drawers of an Antiquarian plan chest will not take either A0 or 60 × 40-in. sheets and so some extra expenditure is necessary on storage furniture.

TABLE 7.1. DRAWING SHEET SIZES

Imperial sizes in inches (metric sizes in millimetres)

Imperial sizes	Metric 'A' series.
	4A0 (2,378 × 1,682)
72 × 40 (1,829 × 1,016)	
	2A0 (1,682 × 1,189)
60 × 40 (1,524 × 1,016)	
53 × 30 (1,346 × 762)	
	A0 (1,189 × 841)
40 × 30 (1,016 × 762)	
40 × 27 (1,016 × 686)	
40 × 15 (1,016 × 381)	
	A1 (841 × 594)
30 × 22 (762 × 559)	
30 × 20 (762 × 508)	
27 × 20 (686 × 508)	
	A2 (594 × 420)
20 × 15 (508 × 381)	
	A3 (420 × 297)
15 × 10 (381 × 254)	
13 × 8 (330 × 203)	
	A4 (297 × 210)
10 × 8 (254 × 203)	
	A5 (210 × 148)
	A6 (148 × 105)

Where drafting film or paper is purchased in roll form the standard roll widths are: 40 in. and 30 in. (1,016 and 762).

TABLE 7.2. DRAWING BOARD AND SHEET SIZES
(Imperial sizes in inches, Metric sizes in millimetres)

Boards	Sheets
	4A0 (2,378 × 1,682)
84 × 58 (2,134 × 1,473)	
84 × 44 (2,134 × 1,118)	
80 × 40 (2,032 × 1,016)	
78 × 48 (1,981 × 1,219)	
	2A0 (1,682 × 1,182)
73 × 40 (1,854 × 1,016)	
62 × 42 (1,575 × 1,067)	
60 × 45 (1,524 × 1,143)	HAMBURG
60 × 40 (1,524 × 1,016)	
54 × 36 (1,372 × 914)	ANTIQUARIAN
	A0 (1,189 × 841)
54 × 32 (1,372 × 813)	
46 × 33 (1,168 × 838)	
42 × 34 (1,067 × 864)	
42 × 33 (1,067 × 838)	DOUBLE ELEPHANT
42 × 32 (1,067 × 813)	
42 × 29 (1,067 × 737)	
	A1 (841 × 594)
32 × 27 (813 × 685)	IMPERIAL
32 × 23 (813 × 584)	
	A2 (594 × 420)
23 × 16 (584 × 406)	HALF IMPERIAL
	A3 (420 × 297)
16 × 12 (406 × 305)	QUARTER IMPERIAL
	A4 (297 × 210)
	A5 (210 × 148)
	A6 (148 × 105)

The sizes A0, A1, A2, A3, A4, A5 and A6 are preferred. The 'A' range of sizes is based on an area of one square metre, the sides of the rectangle being in the ratio $1 : \sqrt{2}$. The sizes are obtained by halving the sheets, folding parallel to the smaller side.

A great many drafting machines in the U.K. and on the Continent are 60 × 40 in. (1,524 × 1,016) or 60 × 45 in. (1,524 × 1,143). The 60 × 45 in. board is known as a Hamburg size.

7.3. Preferred Scales for Drawings

The familiar 'quarter-scale' and 'eighth-scale' are not preferred under the ISO System. The preferred scales are:

1 : 1 1 : 2 1 : 5
1 : 10 1 : 20 1 : 50
1 : 1250
1 : 2500

7.4. Presentation of Toleranced Dimensions

There are four ways of writing a toleranced dimension:

System A:	9.475 (mm in every case)
	9.450
System B:	9.475 – 0.025
System C:	9.500 – 0.025
	– 0.050
System D:	9.500 e8

The significance of the symbols e8 will be discussed in Chapter 8.

Discussion of the systems follows these lines:

System A is an attempt at realism by specifying the size limits to be measured. The operator has no arithmetic to perform.

System B gives the size the designer really wants together with the tolerance away from the desired size.

System C gives a 'nominal size' usually one which appears in a Renard Series or a round number. 9.5 appears in the R40 series.

The appearance of two limits having the same sign in System C is rather awkward but the arrangement has the great advantage of being directly related to the numbers in ISO tables of limits and fits.

The most popular system among draughtsmen and workshop staff is System A. It is said that giving maximum and minimum sizes makes a drawing easier to read especially when soiled or worn. Small plus and minus signs are said to be obscured all too easily under these conditions.

Continental drawings often show System D but British practice is only to give System D in addition to one of the other systems, never alone.

7.5. Introduction of Microfilm Storage of Drawings

The introduction of metric "A" sizes for drawing sheets is a suitable occasion to introduce microfilm storage of drawings. This is because the aspect ratio (that is the ratio of the lengths of long and short sides, $1 : \sqrt{2}$ for A sizes) is constant and is identical to that of the effective area of the film frame.

Those drawing offices which already use microfilm for reproducing, reducing, storing and transmitting drawings will already be aware of the advantages and problems associated with the technique.

Those who do not use microfilm are advised to consider its introduction concurrent with the change to metric practice.

7.6. Advantages of Microfilming

The reduction of the space required for storage of information in the form of engineering drawings is the most obvious advantage of microfilming. The associated reduction in the cost of storage furniture is also well known.

The reduction in fire hazard is a consideration which merits attention. The amount of time and money represented by a stock of original drawings is far too great to be worth the risk of destruction by fire. Microfilm can be stored in enclosures in such a way that the risk of destruction by fire is negligible. The ease with which a drawing may be reproduced to either its original scale, or to a reduced scale, is an advantage.

The problems associated with the handling, storage and display of large drawing sheets are considerable. Many production processes can be planned and carried out with the aid of reduced-scale drawings.

7.7. Problems Associated with Microfilming

The essential requirement is that the finest detail on the original drawing shall be recorded and reproduced without significant loss of quality.

Supplement No. 1 to British Standard 308, *Engineering Drawing Practice*, entitled "Recommendations for the Preparation of Drawings Suitable for Microfilming", follows the principles laid down in BS 308 but emphasizes open-plan presentation and the avoidance of cramped detail and cross-hatching wherever avoidable.

It is also necessary to provide camera alignment marks and a figured scale on an original intended for microfilming.

7.8. Drawing for Microfilming

The detailed recommendations in the Supplement to BS 308 refer particularly to unperforated 35-mm film and cover single frame storage of drawings up to A1 (594 × 841 mm) or 30 × 40-in.

Larger drawings have to be filmed in sections in most cases although equipment can be obtained for single frame filming up to A0 (841 × 1189 mm) or 30 × 53-in.

Drawings should be prepared on matt-surfaced white or translucent drafting material. Non-white drafting materials may be used provided adequate contrast is maintained between line and background.

Pre-printed sheets with the printing on the back may lose quality slightly in reproduction.

Matt surface ink or pencil lines are suitable.

The working surface under the drafting material should be sufficiently hard to avoid excessive indentation under pencil since the scattering of light by indentations can adversely affect clarity of reproduction.

Line density should be as high as possible and as uniform as possible.

Hatching should be omitted wherever possible. If the meaning of the drawing makes hatching unavoidable the lines should not be close together. Areas of fine detail should be shown in separate, enlarged, views.

7.9. Camera-alignment Marks and Scales

In the centre of each of the four sides of the drawing an outline arrow head should be drawn. The point of the arrow should be facing outwards and touching the frame line.

The size of the copy printed back from a microfilm drawing will usually be different from the original. Notes indicating the scale of the original, such as Full Size of $1\frac{1}{2}$-in. to 1 ft, are therefore useless. A figured scale, at least 150 mm long, should be drawn and a note 'DO NOT SCALE' appended.

The exact size of the frame, or outline, of the original may be indicated in a box in one corner of the drawing.

CHAPTER 8

The ISO System of Limits and Fits

8.1. Literature on the System

The origin of the present system is to be found in a publication of the International Standards Association (ISA). The publication was known as Bulletin 25 and the final issue was in 1942. Bulletin 25 contained the metric version of an internationally acceptable system which had been in use on the Continent since about 1932.

In 1946 the ISO took up the work of international standardization and, in 1949, a Technical Committee, ISO/TC3, Limits and Fits, examined the question of revising and modernizing the ISA System. The outcome of the ISO work was the issue, in 1962, of Recommendation 286. This proved to be an extended and corrected version of the ISA System. ISO/R286 contains both inch and metric tables and hence is truly international.

The metric version is reproduced in British Standard Handbook 18: 1966, *Metric Standards for Engineering*.

8.2. International Acceptance of the ISO System

The current British Standard BS 1916: 1953, *Limits and Fits for Engineering*, is actually based on the original ISA system.

BS 1916 is published in three parts:

Part 1: *Limits and Tolerances.*
Part 2: *Guide to the Selection of Fits.*
Part 3: *Recommendations for Tolerances, Limits and Fits for Large Diameters*

Work is in hand to bring BS 1916 into line with ISO/R286. The revisions are, however, minor ones and Part 2 of BS 1916, the 'Guide', is still a valuable text which contains explanatory material on the nomenclature and application of the system.

The American Society of Mechanical Engineers published, in 1942, a version of the English text of ISA Bulletin 25 with a set of tables converted from metric into inch units.

Since the publication of BS 1916, Canada and the U.S.A. have followed the British Standard.

ASA B.41 : 1955 and
CSA B.97 : 1954 (revised 1963)

are in line with BS 1916.

The ISO System is, therefore, the basis of a truly international standard.

The American and Canadian versions use a selection of unilateral hole fits and use a special set of symbols to designate the fits. A translation table is necessary to interpret these in terms of ISO symbols.

8.3. Scope of the ISO System

ISO/R286 offers a modern, comprehensive and systematic set of tolerances, limits and fits amply covering all classes of engineering from the most coarse to gauge and instrument manufacture. The fact that a single set of tables covers production and gauge making is a great advance over the previous BS 164: 1924 which did not specify gauge limits, and none were published until BS 969 appeared in 1941.

It is unlikely that any one organization would ever require to cover the entire range offered, and the intention is that local and particular applications will be covered by selections of preferred class of fit and grade of tolerance from the tables.

8.4. The Structure of the System

The ISO System separates the two fundamental concepts involved in dimensioning parts which are to be members of fit.

The first concept is that of tolerance. Tolerance is the acceptance of an error of workmanship. It is an admission that a part cannot, economically, be made to exactly the size the designer wishes.

The designer has to reconcile his desire to obtain a component of a stipulated size with the fact that severe, or 'tight' grades of tolerance will involve higher production costs than will easy, or coarse grades.

The second concept is that of deviation. Deviation is the algebraic difference between the size required and the basic size. The

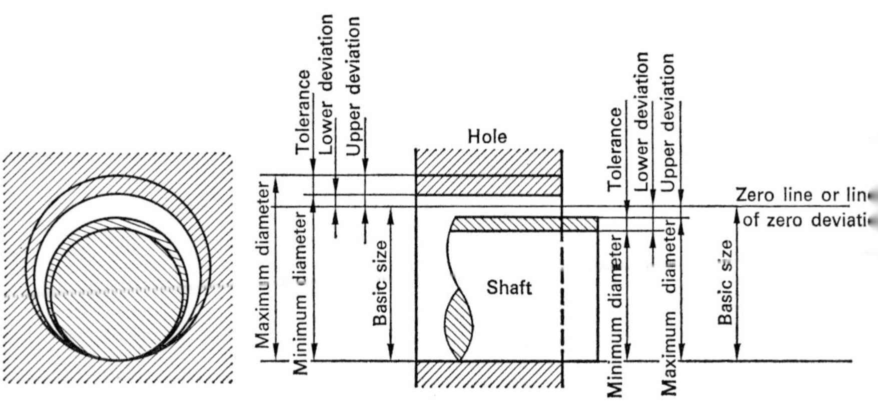

FIG. 8.1. Diagrammatic illustration of definitions in limits and fits. (Reproduced from BS Handbook 18: 1966, p. 15.)

basic size is the same for both members of fit. Deviation is introduced intentionally by the designer in order to obtain the type of fit he requires. If the designer lived and worked in a world of perfect machine tools, manned by perfect operators so that he need never grant any tolerance at all, he would still need to stipulate a deviation of one or both members of fit from the basic size in order that they would fit together in the way he required.

Deviations are specified by letters. Capital or upper-case letters signify HOLES and small or lower-case letters signify shafts. Figure 8.1 indicates the nomenclature which is set out fully in ISO/R286 and in BS Handbook 18: 1966.

The ISO System offers twenty-eight deviations. The letter symbols range from A to ZC, the letter I being omitted. The first ten letters of the alphabet, A to G, specify deviations which tend to produce clearance fits; that is, oversize holes and undersize shafts. Thus an 'A' hole would be a hole larger than the basic size and an 'a' shaft would be a shaft smaller than the basic size.

The letters K to ZC represent deviations tending to produce interference fits; that is, undersize holes and oversize shafts. Thus a 'ZC' hole would be a hole smaller than the basic size and a 'zc' shaft would be a shaft larger than the basic size. Figure 8.3 shows the deviations drawn to scale.

The designer, having specified the basic size of the members of fit, and the deviation of one or both from the basic size, must complete his instructions to the operator by stipulating the error of workmanship which will be acceptable in producing the parts. This is achieved in the ISO nomenclature by quoting a number which specifies a grade of tolerance. The system provides eighteen grades of tolerance. They are known as the IT (for International Tolerance) numbers and range from IT01, IT0, through IT1 to IT16. Low numbers indicate small tolerances, that is to say high-grade, expensive work. The larger the number the more coarse the tolerance grade and so, relatively, the lower the cost of production.

The tables are constructed using basic units of 0.001 mm (or 1 micron) for the metric version and 0.001 in. (or 1 thou) for the Imperial version. The numerical values are related to the basic size as well as to the deviation or tolerance grade concerned. In order to construct the tables in this way a series of nominal diameter steps has been chosen throughout the range and the deviations and tolerance grades are calculated by a formula which includes functions of the nominal diameter.

Referring to Fig. 8.1, it will be seen that all the relevant information on the disposition of tolerance zones relative to the basic size is contained in the top zone of the diagram. Since only the top zone of the diagram is of interest it is customary to show only this zone in illustrations and so the form of Fig. 8.2 is arrived at.

In Fig. 8.2 the disposition of the tolerance zones relative to the zero line is shown greatly enlarged, and the other end of the diameter is taken to be somewhere off the bottom of the figure.

Referring to Fig. 8.2 and examining the right-hand portion, the hole basis fits, we see that a basic hole is one in which the lower deviation is zero. That is to say the smallest hole corresponds to the basic size and only error of workmanship is allowed for. The vertical depth of the shaded portion indicates the size of the tolerance zone.

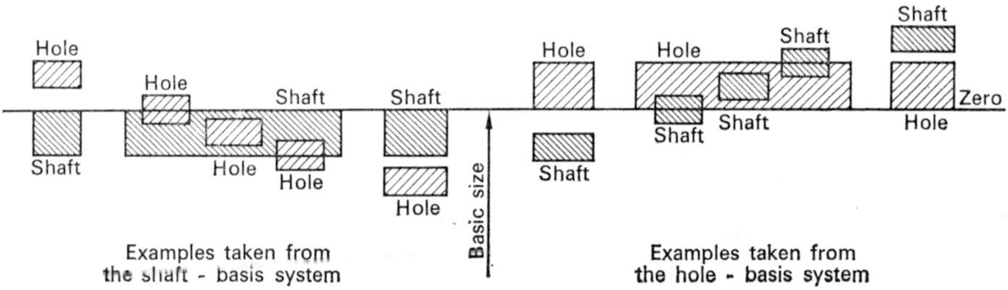

FIG. 8.2. Shaft and HOLE basis systems of fits. (Reproduced from BS Handbook 18: 1966, p. 17.)

The disposition of the tolerance zone relative to the zero line is determined by the deviation and it is this which determines the class of fit.

Three classes of fit are possible: clearance, transition and interference. Referring again to the right-hand half of Fig. 8.2 (the portion to the right of the Basic Size arrow). The left-hand pair of tolerance zones show a clearance fit. The tolerance zone of the shaft lies wholly below that of the hole. We conclude that in this pair of members of fit the largest shaft would always be smaller than the smallest hole. This is the definition of a true clearance fit.

The central block of tolerance zones shows a set of three transition fits. In this condition the type of fit depends on the combination of sizes. A pair of members of fit may have a fit varying from clearance to interference according to selection. It is in this region that the interaction between deviation and tolerance is most clearly seen.

The right-hand pair of tolerance zones shows a condition where the smallest shaft will always be larger than the largest hole. This will result in a true interference or force fit.

All this information is presented by a set of four symbols such as:

H9 — e9

(The hole limits are always placed first).

In which the H tells us that we are dealing with a hole basis pair. The figure 9 sets the grade of tolerance and is in this case, but not in all cases, the same for both members. The letter e tells us that we are dealing with an undersize shaft, therefore with a clearance fit.

Reference to Fig. 8.3 will aid the visualization of the types of fit obtainable. It is obvious that the number of sensible combinations will be limited. Many combinations would be ridiculous. This conclusion leads us on to the next consideration, that of selections of fits.

Figure 8.4 shows ten pairs which cover the vast majority of normal engineering requirements. In Fig. 8.5 the selected fits are drawn to scale and labelled to show the classes of fit obtained.

The symmetrical bilateral tolerance marked js in Fig. 8.3 is available if required but does not appear to be in common use.

In special branches of engineering, such as ball-bearing manufacture, it is more convenient to have a selection of shaft basis fits. Figures 8.6 and 8.7 show the most used selection.

Thus we have seen how practically any conceivable class of fit can be called up by a set of four symbols in two pairs, one pair giving the deviation and tolerance grade for the hole and the other pair giving the deviation and tolerance grade for the shaft.

One interesting consequence of the manner of presentation is that if a shaft-basis fit is required of exactly the same class as a

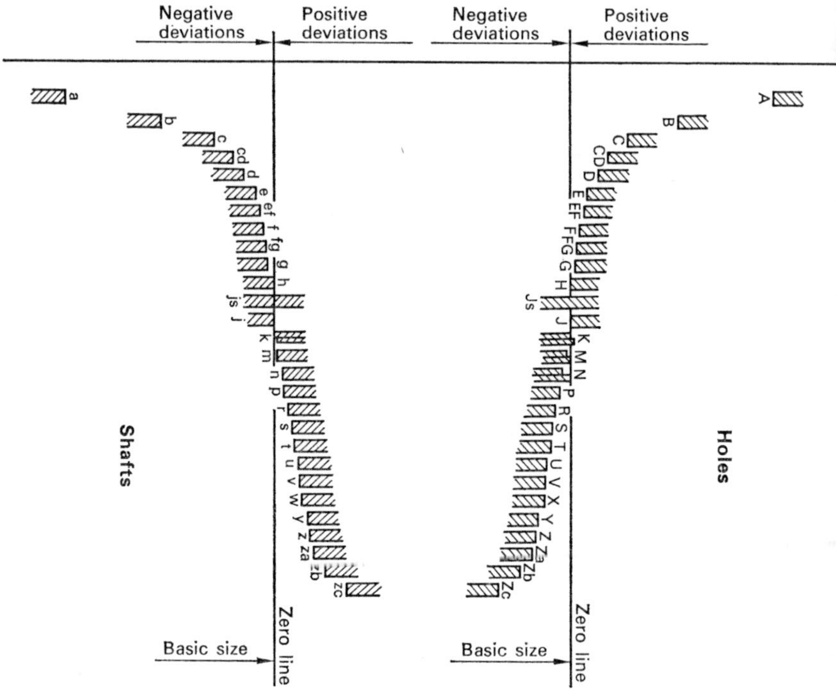

FIG. 8.3. Respective positions of the various tolerance zones for a given diameter step. (Reproduced from BS Handbook 18: 1966, p. 18.)

known hole-basis fit then all that is necessary is a reversal of the letters, leaving the numbers unchanged, thus.

H8 — f7

change to F8 — h7 which is also one of the selected fits and will be of the same class as the original.

8.5. Conversion Accuracy

Now it is necessary to reintroduce a topic discussed in Chapter 6, the conversion from inch to millimetre dimensions or vice versa.

Type of fit	Shaft tolerance	Hole tolerances			
		H7	H8	H9	H11
Clearance	c11				▨
	d10			▨	
	e9			▨	
	f7		▨		
	g6	▨			
	h6	▨			
Transition	k6	▨			
	n6	▨			
Interference	p6	▨			
	s6	▨			

FIG. 8.4. Selected hole basis fits for general engineering. (Reproduced from BS Handbook 18: 1966, p. 36.)

In Chapter 6 we noted how, if the rules given in BS 2856 were followed then it was all too easy to alter the implication in the expressed tolerance. In many simple cases the modified rule given in Chapter 6 will give sufficiently accurate results at reasonable cost. When one moves into the realm of classified fits and tolerances the rigorous methods advocated in BS 2856 will help to minimize the number of cases in which conversion results in a shift into a neighbouring deviation or tolerance. It is, however, quite proper to investigate the relative claims of economics and precision and to divide products into two classes, one in which

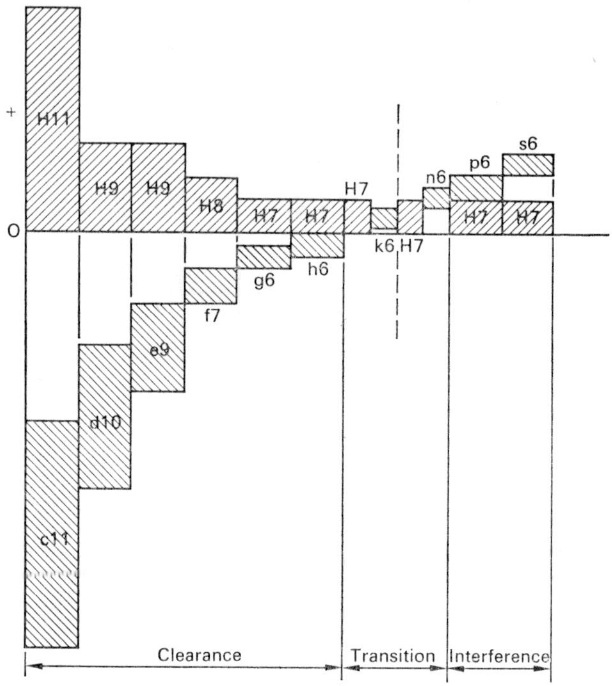

FIG. 8.5. Selected hole basis fits drawn to scale.
(Reproduced from BS Handbook 18: 1966, p. 37.)

class of fit is of paramount importance and one in which production cost is the vital factor.

Consider an example.

A shaft is dimensioned thus

$$\begin{array}{l}2.374 \text{ inches}\\ 2.373\end{array}$$

We could write this as

$$\begin{array}{l}2.375 \text{ nominal } (2\tfrac{3}{8})\\ -0.001\\ -0.002\end{array}$$

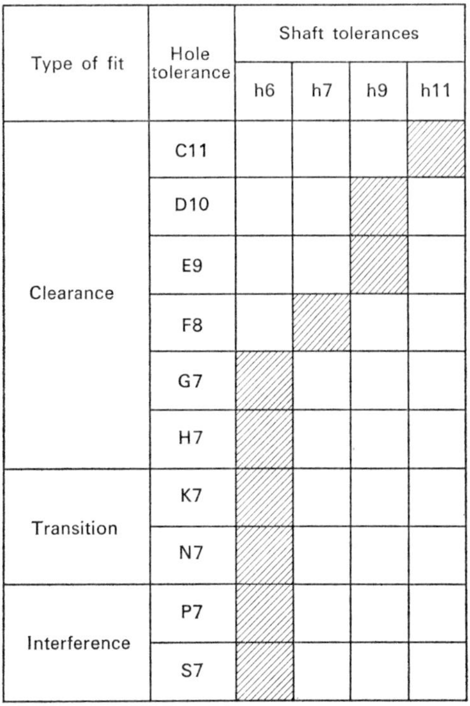

FIG. 8.6. Selected shaft basis fits for general engineering. (Reproduced from BS Handbook 18: 1966, p. 39.)

The tolerance is 0.001.
The deviation is 0.001.

The nearest tabulated values in ISO R286 are:

For tolerance grade (0.001)

IT6	0.000 7	30% too severe
IT7	0.001 2	20% too easy.

For upper deviation es (– 0.001)

f	– 0.001 2	20% too large.

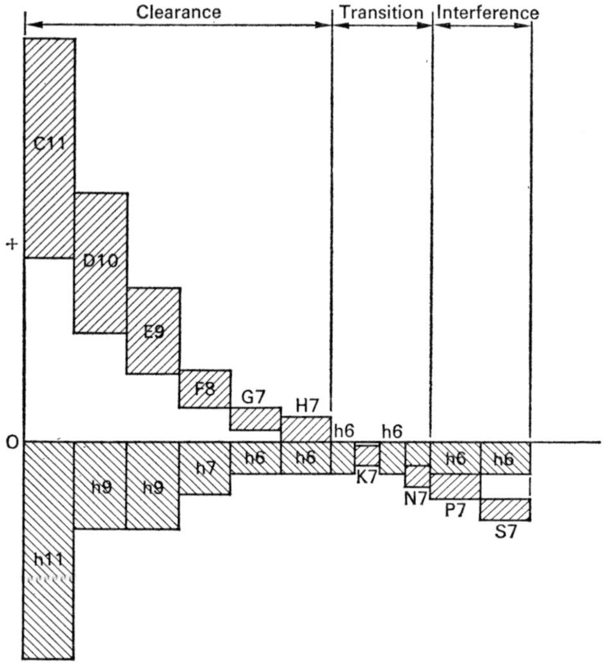

FIG. 8.7. Selected shaft basis fits drawn to scale. (Reproduced from BS Handbook 18: 1966, p. 40.)

Now convert, according to the rules in BS 2856,

$$\begin{aligned} 2.375 \text{ in.} &= 60.325 \text{ mm} \\ 2.374 \quad &= 60.300 \\ 2.373 \quad &= 60.274 \end{aligned}$$

The accompanying table (8.1) shows the values obtained by converting the toleranced dimension by two methods:

(a) By direct conversion of the values given.

(b) By classifying the fit by reference to tolerance and deviation tables and applying the same classification to the converted values.

It will be seen that the fit as given cannot be expressed in ISO system terminology without introducing a 20% error in the upper deviation and a 24% error in the tolerance which the designer originally called for. It would be necessary to establish exactly what the designer's intentions were, and what adjustments would be permissible, before setting down the metric version.

TABLE 8.1.
TOLERANCE COMPATIBILITY

Dimension	Value as given in inches	Value obtained by direct conversion to mm	Value obtained by assuming nearest tabulated class of fit, inches	Value obtained by extracting corresponding items from tables, mm	Percentage error %
Nominal size	2.375	60.325	2.375 f6	60.325 f6	—
Maximum limit of size	2.374	60.300	2.3738	60.2945	0.009
Minimum limit of size	2.373	60.274	2.3731	60.2765	0.004
Upper deviation es	−0.001	−0.025	−0.0012	−0.030	20
Lower deviation ei	−0.002	−0.051	−0.0019	−0.049	4
Tolerance	0.001	0.025	0.0007	−0.019	24

CHAPTER 9

The ISO Metric Screw-thread System

9.1 The Need for Rationalization

Of all branches of engineering it is the design and manufacture of threaded fasteners which most urgently require rationalization. There are some 210 different screw-thread systems in the world, ten of these being metric screw threads in common use.

The Whitworth form was, and still is, widely accepted throughout Europe and much of the rest of the world. However, because of its basis on fractional inch dimensions, the Whitworth thread failed to maintain its position and many metric countries developed their own millimetre-based series.

The United States of America developed its own American National Thread and as the usage of American Coarse, American Fine and the various metric threads began to spread it became clear that the Whitworth thread had lost the day.

The next attempts at standardization took place quite independently, one in metric Europe with the announcement of the 'System International' metric thread, and one in the inch-using world with the 'Unified Thread' series.

Unified threads were introduced in 1949 by the ABC countries: America, Britain and Canada. In their 20 years of existence Unified threads have only managed to secure about a quarter of the market. The attempt to rationalize may therefore be said to have failed and Unified threaded fasteners constitute merely one more line in an already excessively diverse range.

A recent (1968) survey by Japan of the world usage of triangular screw threads showed that, of the eighteen nations questioned, sixteen nations intended to make ISO Metric threads the primary system for general purposes. One nation (U.S.A.) intended to operate a dual system. (The eighteenth nation, New Zealand, was undecided.) The term ISO inch is the name for Unified threads. In 1953 the International Standards Organization decided to recommend the revised Unified Thread System as the acceptable inch-based international system. Since that time it has been known as the ISO Inch System.

9.2. ISO Metric Screw Threads

The ISO Recommendations numbered 261 and 262, issued in 1962, are intended to supersede the ISA System. The series is very similar to the SI and DIN (German) metric screw thread series, but there are tolerance grades in which interference could occur between ISO and SI/DIN components. Figure 9.1 shows the basic

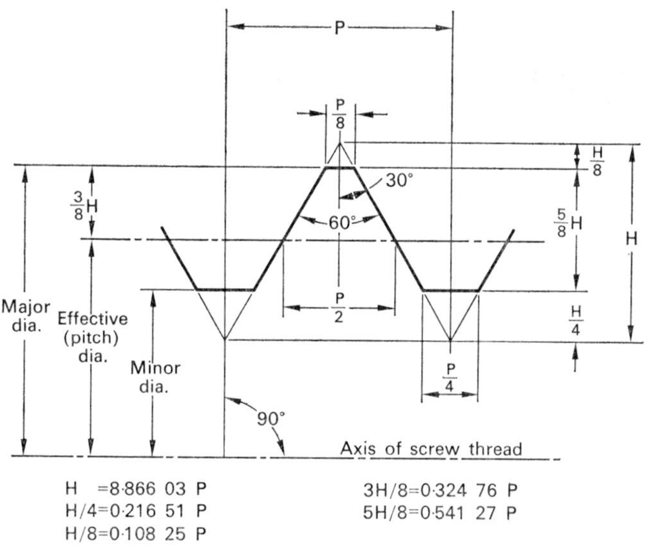

FIG. 9.1. Basic form of ISO thread.
(Reproduced from BS Handbook 18: 1966, p. 81.)

form of ISO Metric thread. The proportions are derived from the relations (see also Fig. 9.3):

Thread angle (2θ)	= 60 degrees
Flank angle (θ)	= 30 degrees
Triangular height H	= 0.866 03 p
Truncation of crest of screw	= H/8
Truncation of crest of nut	= H/4
Truncation of root of screw	= H/6
Radius at root of screw (r_1)	= H/6
Truncation of root of nut	= H/12
Radius at root of nut (r_2)	= H/12
Depth of thread in screw (h_s)	= (17/24) H
Depth of engagement (h)	= (5/8) H
	= 0.541 27 p
Major diameter of screw (D)	= nominal diameter
Effective diameter (E)	= D – 0.649 52 p
Minor diameter of screw (d_s)	= D – 1.226 87 p
Minor diameter of nut (d_n)	= D – 1.082 53 p

It will be apparent that this profile is identical with that of the ISO Inch (unified) thread. Since the ISO Metric profile will be applied to fasteners whose nominal diameters are round or preferred numbers of millimetres; and the ISO Inch profile will be applied to fasteners whose nominal diameters are round or preferred fractional inches; the two series of fasteners will not be interchangeable. The tolerance systems for the two series are not identical. It was decided in 1964 that, since the items are not interchangeable anyway, there was no point in expending time and money on bringing the tolerance systems into line.

The 'First Choice' nominal diameters follow an irregularly rounded R10 Renard Series. There are three series in the general plan: a coarse pitch for general purposes, and fine and constant pitch for special purposes. Only the coarse series is recommended for normal use. Let us consider the question of pitch/diameter relationships with the aid of Fig. 9.2. The figure shows nominal diameter plotted horizontally against pitch plotted vertically for various widely used screw threads. It will be seen that the more coarse-pitched a thread the steeper the line will be. The manner in which the fastener will fail is, to some extent, determined by the pitch/diameter ratio. Very fine pitch threads, that is shallow lines on the chart, will tend to fail by stripping. Very coarse threads, that is steep lines on the chart, will tend to fail by core fracture.

BS 3580: 1964, *Guide to Design Considerations on the Strength of Screw Threads*, contains a detailed discussion of modes of failure and discusses thread bending, nut bursting, the determination of the correct ratio of nut strength to bolt strength and other relevant topics.

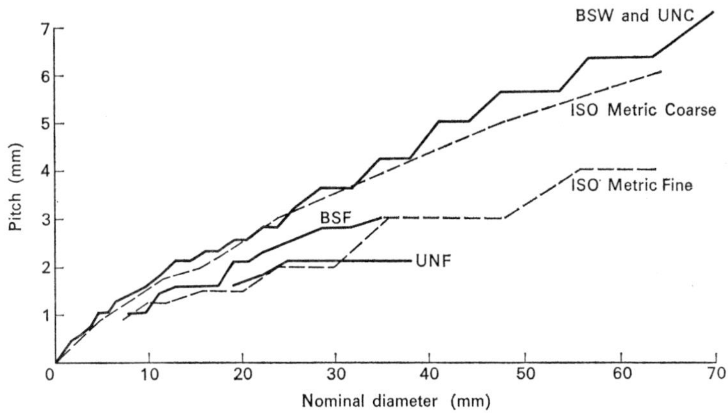

FIG. 9.2. An analysis of some screw thread systems.

For our immediate purpose, however, all we need to note is that the Whitworth thread series (BSW) is sufficiently coarse to ensure bolt fracture rather than thread stripping and that the ISO Metric Coarse and Unified Coarse series are very close to BSW. Unified Fine (UNF) have been little used above 1 inch (25 mm) diameter and the horizontal line marked UNF does not represent a series in practical use.

From the chart it is apparent that the replacement of BSW by ISOM or UNC is a safe and practical proposition.

At this point a word of explanation should be given of the use of the symbols (ISOM) for ISO Metric Coarse. Since ISO Metric Coarse is the only series recommended for normal use any reference such as 'ISO Metric', 'ISOM' or "Metric" is taken to refer to the coarse series. The nominal diameter in millimetres is preceded by the letter 'M' thus; M20 means a 20-mm nominal diameter ISO Metric coarse thread.

When it is necessary to specify any other thread, fine or constant pitch, the pitch in mm must be stated thus: M20 × 2.5 means a 20-mm fine-pitch ISO Metric thread. The question of indicating tolerance grade and position will be dealt with later.

ISO Metric Fine threads do not seem to be finding much favour in Britain. The manufacturers of large diesel engines regard it as a stripping thread and one which all too easily gets cross-threaded. One reason for its greater usage in Europe may be that ISO Limits and Fits are more correctly applied by European users. This brings us to a discussion of tolerances for screw threads, which will include the subject of root and crest rounding.

9.3. Design Form and Tolerances

In developing, from the basic profile, the design profiles, two key factors have to be considered. Contact between the numbers of fit has to be restricted to the flank areas, over a radial depth of 5H/8; and interference between roots and crests has to be avoided. These conditions must still apply in the extreme case when both members of fit are in their maximum metal condition, that is to say a maximum size bolt mating with a minimum size nut. This extreme case sets the limits to the amount of rounding of roots which can be permitted.

Smooth rounding of thread roots is a matter of importance, since sharp corners at the root are serious stress raisers. Rounding of crests is not of importance since the crest is unstressed. Various forms of crest are permissible, so long as the profile remains within the limits.

Figure 9.3 shows the design form of the external thread. The important feature, as stated, is the root radius.

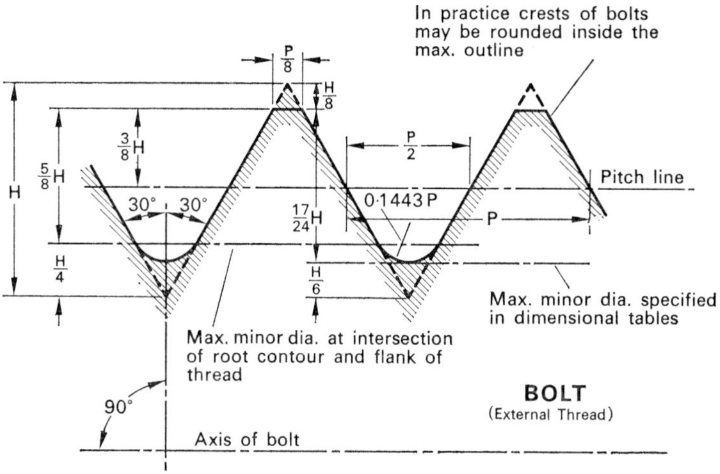

FIG. 9.3. Design form of ISO metric thread (external). (Reproduced from BS Handbook 18: 1966 p. 83.)

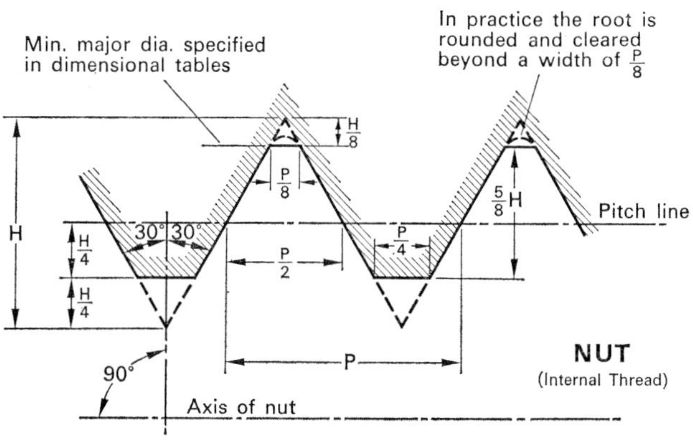

FIG. 9.4. Design of ISO metric thread (internal). (Reproduced from BS Handbook 18: 1966, p. 82.)

Figure 9.4 shows the other member of fit, the internal thread. Here again the root rounding will be noticed.

The idea of the location of tolerance elements may be introduced with the aid of Fig. 9.5.

As in the case of limits and fits for plain work the idea of grade of tolerance may be treated separately from the deviation although the inter-action between them has to be considered.

ISOM fasteners may be made to three classes of fit designated Fine, Medium and Coarse. Fine class should only be used for precision work where cost is not a primary consideration.

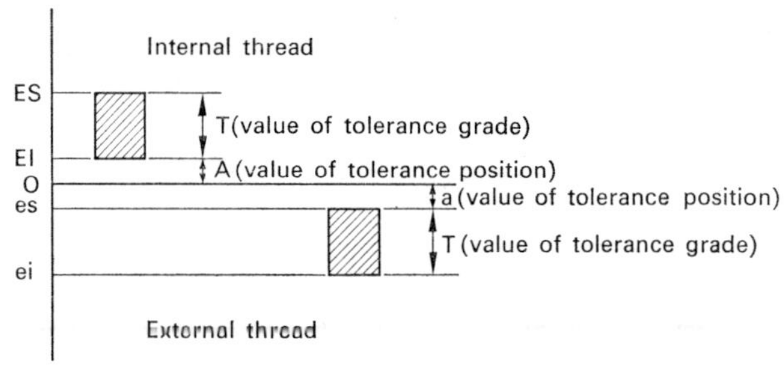

FIG. 9.5. Location of screw thread tolerance elements. (Reproduced from BS Handbook 18: 1966, p. 89).

Medium is the general commercial grade and is, in fact, what customers receive when they omit to specify any tolerance requirement at all (some 55% of orders received by one firm for threaded fasteners do not stipulate a class of fit).

Coarse is a grade for use under special conditions such as dirty environments.

It is unfortunate that the words coarse and fine are applied to both pitch and class of fit, care must be taken not to confuse the ideas.

As in the plain limits and fits system capital letters are used to denote the deviations of nuts, small letters denote the deviations of bolts. A class of fit is specified by four symbols, a pair made up of a capital letter and a number for the deviation and tolerance grade of the nut, and a pair made up of a small letter and a number for the deviation and tolerance grade of the bolt. Normal class is 6H–6g.

Table 9.1 shows the choice of classes for normal length of engagement, and uncoated threads. Where long engagement or coatings are applied to the threads special classes are stipulated in BS 3643: 1963, *ISO Metric Screw Threads* and in BS Handbook 18: 1966, *Metric Standards for Engineering.*

TABLE 9.1 TOLERANCE CLASSES FOR NORMAL LENGTHS OF ENGAGEMENT

	Tolerance class	
Class of fit	Bolt	Nut
Fine	4h	5H
Medium	6g	6H
Coarse	8g	7H

Figure 9.6 shows the tolerance zones drawn in their correct scale positions on the ISOM design profiles.

A shorthand form for showing a wide range of combinations of deviation and grade is the chart shown in Fig. 9.7. The chain dotted line represents the zero line, corresponding to the basic size. Vertical solid lines represent the extent of the tolerance zones for the numbers indicated. Deviations are indicated by the space between the zero line and the inner end of the tolerance line.

Finally the question of interference between SI and ISO threads: BS 1095: 1943, *Metric Screw Threads, System International,* is withdrawn but gauges, profiles and components produced to comply with it may occasionally be encountered. In the highly unlikely event of a maximum material SI nut finding its way into a batch of ISOM nuts, and in the further unlikely event of that particular nut being partnered with a maximum material ISOM bolt, then interference could occur between nut thread crests and bolt thread roots.

Figure 9.8 indicates the zone of interference. SI forms, in any case, are obsolete since the withdrawal of BS 1095.

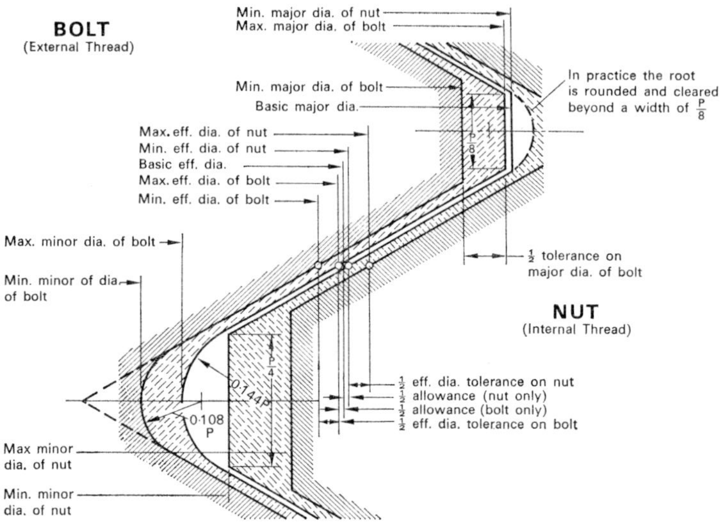

FIG. 9.6. ISO metric thread with allowance for normal purposes. Tolerance positions g (bolt) and H (nut)
(Reproduced from BS Handbook 18: 1966, p. 97.)

9.4. The Strength Grading of High-tensile Fasteners

The existing convention regarding the marking of high-tensile fasteners is based on the metric technical unit of stress, the kilogramme-force per square millimetre. The convention can be used with a very slight alteration in the strength of the material if and when either the MN/m^2 or the hectobar becomes the customary stress unit.

To deal first of all with the situation as it exists in mid-1968:

The bolt or screw head will carry three markings; one 'ISOM' to indicate that the thread is ISO Metric; one to indicate the manufacturer; and one showing two digits, for example 88. The first digit represents one-tenth of the tensile strength in kgf/mm^2. Thus the first digit in the example would mean that the minimum tensile strength of the material of the bolt was 80 kgf/mm^2. The second

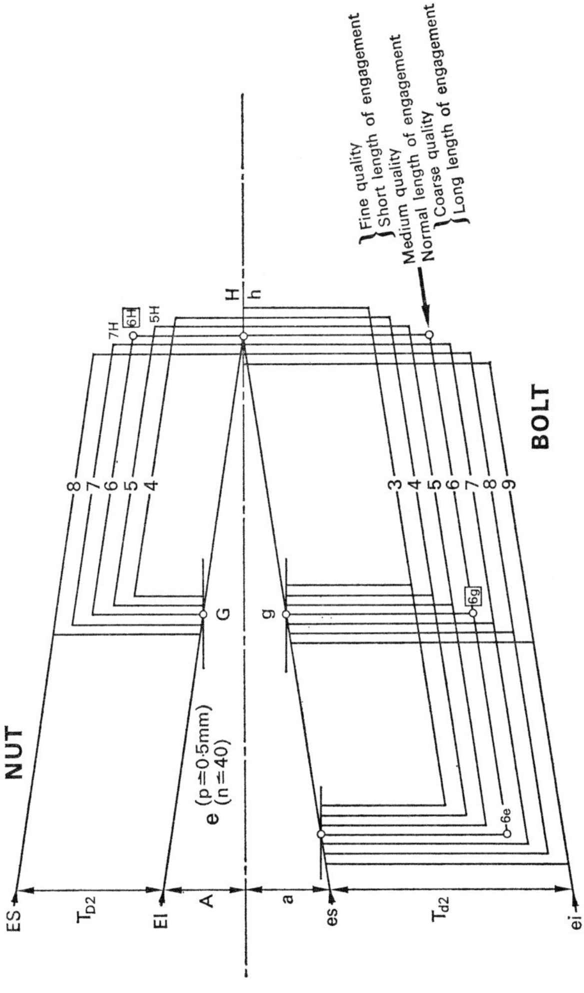

FIG. 9.7. Pitch diameter tolerances for ISO metric screw threads.

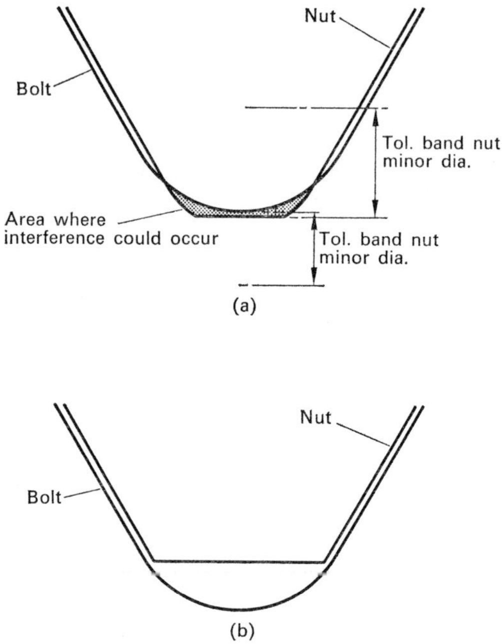

(a)—DIN 13 Blatt 15 nut, ISO metric class 6g bolt, both maximum metal.

(b)—B.S. 1095 SI thread nut, ISO metric class 6g bolt, both maximum metal.

FIG. 9.8. Possible interference between SI nut and ISO bolt. (Reproduced from *The Engineer*, 16 Feb. 1968, p. 279, by permission of the Editor.)

digit indicates ten times the ratio yield strength/tensile strength, in the example 0.8. Thus the yield strength of the bolt material would be $80 \times 0.8 = 64$ kgf/mm^2.

9.5. The Marking of Suitable Nuts

The nut to correspond with the above bolt will carry two marks: one mark, an indented circular dot, is placed at the position corresponding to 12 o'clock on a clock face; the other mark, an indented radial bar, is then in the position corresponding to the number on the clock face which indicates the grade.

A suitable nut for use with an 88 bolt would be grade 8 and the indented radial bar would appear at 8 o'clock.

The nut suitable for use with a high-tensile bolt is softer than the bolt. The proof-load stress of the nut should be equal to the minimum tensile strength of the highest grade of bolt with which the nut is to be used. The number represented by the mark on the nut is one-tenth of the specified proof-load stress.

Thus for a grade 88 bolt we have a tensile strength of 80 kgf/mm^2. The nut proof-load stress would be 80 kgf/mm^2 and the nut marked to correspond to grade 8, that is the bar should be at 8 o'clock.

However, the hectobar has been used as the stress unit in the revised British Standards dealing with the strength of materials. The basic standard in this category is BS 18: *Methods for Tensile Testing of Metals*. It is reasonable to suppose that, if the hectobar becomes the customary stress unit in a matter as fundamental as the strength of metals, then it will infiltrate through all metallurgical, engineering, and structural work.

Should this be the case, the close similarity between the sizes of the units: kgf/mm^2 and hbar will be an advantage.

Since 1 hbar = 1.019 720 kgf/mm^2

or 1 kgf/mm^2 = 0.980 665 hbar

there is less than 2% difference between actual and assumed strengths if the incorrect unit is assumed. If the MN/m^2 is used, the rule becomes 'the first digit represents 100th of the tensile strength in MN/m^2'.

For example: suppose that a bolt marked 88 has, in accordance with ISO/DR 911, *Mechanical Properties of Fasteners*, and BS 3692: 1967, *ISO Metric Precision Hexagon Bolts, Screws and Nuts*, a minimum tensile strength of 80 kgf/mm^2,

80 kgf/mm^2 is 78.6 hbar

and 80 hbar is 81.6 kgf/mm^2.

During the changeover period from metric technical units to SI units there would be a possibility of a user receiving a fastener manufactured from a material having a minimum tensile strength 2% less than the value he assumes.

This situation would, of course, only arise during the period when stocks of fasteners manufactured to the 80 kgf/mm^2 specification lasted.

The alloys and manufacturing processes could be adjusted to increase the strength slightly to comply with the 80 hbar specification.

The issue of technical data brochures on high-tensile fasteners will be the most direct method of informing users of any change. For most applications, of course, 2% is an insignificant difference, and far less than other variations and errors arising in practice.

The vast majority of users could assume either unit without significant risk.

CHAPTER 10

Currency Decimalization

10.1. The Retention of the Pound Sterling

The pound sterling, divided into twenty shillings or 240 pence, is the only non-decimal currency remaining in the world. All other currencies which have subdivisions use units of one-hundredth or one-thousandth of the unit of exchange.

On 15th February 1971 Britain will change to a decimal currency, introducing a new penny, equal to one-hundredth part of the pound sterling.

The retention of the pound sterling as the unit of exchange was regarded by the British Government as a necessary prerequisite for decimalization because of sterling's function as a reserve currency and a medium of exchange for international trade.

10.2. The New Coinage

Given that the pound is to remain; the new penny, 1% of a pound, is a rather coarse unit for fixing the price of low-cost articles. This has led to introduction of a half new penny. Thus we have

$$\begin{aligned} &\text{one new penny} = 2.4 \text{ old pennies} \\ \text{and} \quad &\text{half new penny} = 1.2 \text{ old pennies.} \end{aligned}$$

There will, of course, be a requirement for new coins. The multiples chosen for embodiment in coins are:

$\frac{1}{2}$ new penny = £0.00$\frac{1}{2}$ = $\frac{1}{2}$p
1 new penny = £0.01 = 1p
2 new pence = £0.02 = 2p
5 new pence = £0.05 = 5p
10 new pence = £0.10 = 10p
50 new pence = £0.50 = 50p

The letter p for new penny avoids confusion with the old 240-to-the-pound unit for which the symbol was *d*.

Table 10.1 at the end of this chapter shows the conversions between the pound new penny units and the old £ *s*. *d*. units.

It will be seen that old currency units of 3*d*., 5*d*., 7*d*., 10*d*. and 1*s*. are convenient for conversion into decimal pence. The conversion of amounts other than round numbers of shillings will, however, be seen to involve errors of up to 25%.

1p = 2.4*d*. ≑ 3*d*. within 25% error
2p = 4.8*d*. ≑ 5*d*. ,, 4.2% ,,
3p = 7.2*d*. ≑ 7*d*. ,, 2.8% ,,
4p = 9.6*d*. ≑ 10*d*. ,, 4.2% ,,
5p = 1*s*. exactly

The great advantage of a currency unit decimally subdivided is that normal arithmetical calculating devices can be used. Computers, desk calculators and slide-rules are directly applicable without having to convert shillings and pence into decimals of a pound before embarking on the calculations. When calculating devices are used the new half penny may be written as £0.005 instead of as £0.00$\frac{1}{2}$.

The reason for not showing the new half penny as a third place of decimals in other cases is that the half penny is the last subdivision of the currency and does not, for instance, represent 'five mils'. There are no coins to represent other multiples of one-thousandth of a pound.

A possible source of confusion when writing the calculated cost per article of low-cost articles to decimals of a new penny is the temptation to write, for example, to represent an article costing 2.36 new pence, the symbols 2.36p. This would be wrong as it is liable to be misread as two pounds and thirty-six new pence. In such cases it is essential to write £0.0236.

The amount represented could be rounded to 2p with an error of 18% or to $2\frac{1}{2}$p with an error of 6%.

Rounding down, of course, benefits the purchaser and rounding up benefits the vendor.

10.3. International Exchange Rates

Calculation of international exchange rates is greatly simplified by decimalization. To take the example of the U.S. dollar which stood, on 18th March 1968, at $2.40\frac{1}{4}$ to the pound, we have:

$$\$2.4025 = £1$$

therefore

$$1\$ = \frac{£1}{2.4025} = £0.416$$

or, to the nearest $\frac{1}{2}$p,

$$1\$ = £0.41\tfrac{1}{2}$$

Table 10.2 shows the rates of exchange for most world currencies as they stood on 18th March 1968. The rates are to slide-rule accuracy and should, in any case, be checked against the current issue of the *Financial Times* or by reference to an authorized dealer in the currencies concerned.

10.4. Metrication and Decimalization

Since "D-Day" for currency is 15th February 1971 and since the metrication programme for British industry commences in 1970 and runs through to 1975, it is inevitable that the retail trades will eventually go metric, and the two changes will interact. The changes in legislation necessary for retail metrication are to be accomplished by enabling legislation having an appointed day not later than January 1971.

Let us consider first the everyday problem of purchasing goods by weight. The unit by which goods are bought and sold after metrication of retail trades will be the kilogramme weight. Let us examine the problem of converting

old pence per pound weight
into new pence per kilogramme weight.

Now, since 1p = 2.4*d.*,

we have $$1d. = \frac{1}{2.4}\,\text{p} = 0.41\text{p}$$

and since $$1\ \text{kg} = 2.204\ \text{lb}$$

we have $$1\ \text{lb} = \frac{1}{2.204\ 6}\ \text{kg} = 0.453\ 6\ \text{kg}$$

so that $$\frac{1d.}{1\ \text{lb}} = \frac{0.416}{0.454}\,\frac{\text{p}}{\text{kg}}$$

$$= 0.916\,\frac{\text{p}}{\text{kg}};$$

calling this 1 p/kg involves an error of 9% in favour of the vendor. We conclude therefore that

$$\frac{1d.}{\text{lb}} \doteqdot \frac{1\text{p}}{\text{kg}}$$

within 9%

Another common conversion problem will be the cost of transport in old pence per mile.

Let us convert old pence per mile into new pence per kilometre:

$$\frac{1d.}{\text{mile}} = \frac{0.416}{1.609\ 34}\,\frac{\text{p}}{\text{km}}$$

$$= 0.26\,\frac{\text{p}}{\text{km}}.$$

Thus we should have:

$$\frac{9d.}{\text{mile}} = 9 \times 0.26\,\frac{\text{p}}{\text{km}} = 2.34\,\frac{\text{p}}{\text{km}}.$$

We would be justified in rounding this figure up to

$$2\tfrac{1}{2}\,\frac{\text{p}}{\text{km}} \text{ with an error of } 7\%.$$

Some other common values would be:

$$\frac{6d.}{\text{mile}} = 1.56\ \frac{\text{p}}{\text{km}} \doteqdot \frac{2\text{p}}{\text{km}} \quad \text{error } 28\%,$$

$$\frac{7\frac{1}{2}d.}{\text{mile}} = 1.95\ \frac{\text{p}}{\text{km}} \doteqdot \frac{2\text{p}}{\text{km}} \quad \text{error } 2\tfrac{1}{2}\%,$$

$$\frac{1s.}{\text{mile}} = 3.12\ \frac{\text{p}}{\text{km}} \doteqdot \frac{3\text{p}}{\text{km}} \quad \text{error } 4\%.$$

10.5. Special Problems During Changeover

Coexistence of decimal coinage and traditional coinage during the period 1969 to 1971 will mean that a certain facility in converting mentally from one system to another will be required. A few simple and easily memorized rules are given here:

Perhaps the most useful fact to remember is that new pence are percentages of one pound: 50p is 50% of a pound (i.e.: 10*s*.) 25p is 25% of a pound, i.e. 5*s*., and so on.

The next most useful device is to read, let us say 3*s*. 6*d*. as '36' and halve the number, giving 18, answer 18p. Similarly, 7*s*. 4*d*., say 74, halve it 37, answer 37p and so on.

The method works equally well in reverse; 72p double it, 144 answer 14*s*. 4*d*. Similarly, 32p, double it 64 answer 6*s*. 4*d*.

An extension of this problem is that change may have to be calculated in either or both systems.

Example: I enter a shop to purchase an article priced at 3*s*. 6*d*. I proffer two 10p pieces, how much change should I receive?

Two 10p pieces total 20p . . . double it 40, total proffered is therefore 4*s*. 0*d*. Change is 6*d*.

Second example: I wish to purchase an article priced at 3*s*. and I proffer one 10p and one 5p piece. How much change should I receive?

10p, double it, 20, say 2*s*. 0*d*.
5p, double it, 10, say 1*s*. 0*d*.
3*s*. 0*d*.

Therefore I shall not receive any change. 15p is exactly 3*s*. 0*d*.

At first only the 5p and 10p pieces will circulate and; since they are equivalent to 1*s*. and 2*s*., this keeps the above problems simple.

TABLE 10.1. CONVERSION CHART: DECIMAL CURRENCY TO £ *s. d.*

£ np	£ s. d.	Round nos. in £ s. d.	£ np	£ s. d.	Round nos. in £ s. d.
0.005	0 0 1.2		0.37	0 7 4.8	
0.01	0 0 2.4	3	0.38	0 7 7.2	
0.02	0 0 4.8	5	0.39	0 7 9.6	
0.03	0 0 7.2	7	0.40	0 8 0	0 8 0
0.04	0 0 9.6	10	0.41	0 8 2.4	
0.05	0 1 0	0 1 0	0.42	0 8 4.8	
0.06	0 1 2.4		0.43	0 8 7.2	
0.07	0 1 4.8		0.44	0 8 9.6	
0.08	0 1 7.2		0.45	0 9 0	0 9 0
0.09	0 1 9.6		0.46	0 9 2.4	
0.10	0 2 0	0 2 0	0.47	0 9 4.8	
0.11	0 2 2.4		0.48	0 9 7.2	
0.12	0 2 4.8		0.49	0 9 9.6	
0.13	0 2 7.2		0.50	0 10 0	0 10 0
0.14	0 2 9.6		0.51	0 10 2.4	
0.15	0 3 0	0 3 0	0.52	0 10 4.8	
0.16	0 3 2.4		0.53	0 10 7.2	
0.17	0 3 4.8		0.54	0 10 9.6	
0.18	0 3 7.2		0.55	0 11 0	0 11 0
0.19	0 3 9.6		0.56	0 11 2.4	
0.20	0 4 0	0 4 0	0.57	0 11 4.8	
0.21	0 4 2.4		0.58	0 11 7.2	
0.22	0 4 4.8		0.59	0 11 9.6	
0.23	0 4 7.2		0.60	0 12 0	0 12 0
0.24	0 4 9.6		0.61	0 12 2.4	
0.25	0 5 0	0 5 0	0.62	0 12 4.8	
0.26	0 5 2.4		0.63	0 12 7.2	
0.27	0 5 4.8		0.64	0 12 9.6	
0.28	0 5 7.2		0.65	0 13 0	0 13 0
0.29	0 5 9.6		0.66	0 13 2.4	
0.30	0 6 0	0 6 0	0.67	0 13 4.8	
0.31	0 6 2.4		0.68	0 13 7.2	
0.32	0 6 4.8		0.69	0 13 9.6	
0.33	0 6 7.2		0.70	0 14 0	0 14 0
0.34	0 6 9.6		0.71	0 14 2.4	
0.35	0 7 0	0 7 0	0.72	0 14 4.8	
0.36	0 7 2.4		0.73	0 14 7.2	

TABLE 10.1 (*cont.*)

£ np	£ s. d.	Round nos. in £ s. d.	£ np	£ s. d.	Round nos. in £ s. d.
0.74	0 14 9.6		0.87	0 17 4.8	
0.75	0 15 0	0 15 0	0.88	0 17 7.2	
0.76	0 15 2.4		0.89	0 17 9.6	
0.77	0 15 4.8		0.90	0 18 0	0 18 0
0.78	0 15 7.2		0.91	0 18 2.4	
0.79	0 15 9.6		0.92	0 18 4.8	
0.80	0 16 0	0 16 0	0.93	0 18 7.2	
0.81	0 16 2.4		0.94	0 18 9.6	
0.82	0 16 4.8		0.95	0 19 0	0 19 0
0.83	0 16 7.2		0.96	0 19 2.4	
0.84	0 16 9.6		0.97	0 19 4.8	
0.85	0 17 0	0 17 0	0.98	0 19 7.2	
0.86	0 17 2.4		0.99	0 19 9.6	

TABLE 10.2. UNITS AND CONVERSION FACTORS FOR INTERNATIONAL TRADE
Exchange Rates are Official Values on 18th March 1968 (*Financial Times*)

State or Nation		Meast. system Date of adoption		Currency system	Conversion factor to sterling		Conversion factor into decimal currency	Nearest round figure in decimal currency
					'To the pound'	Value of unit in pence		
Aden		Imp						
Afghanistan		M	1920	Afghani of 100 puls	181		1 Afghani = £0.005 52	0½p
Albania	ISO	M	1950	Lek of 100 qintar (New Lek)	12		1 Lek = £0.083 3	8p
Algeria		M	1943	Dinar of 100 centimes	11.80		1 Dinar = £0.084 7	8½p
Andorra				Spanish Peseta of 100 centimos	167⅛		1 Peseta = £0.005 98	½p
Angola				Escudo of 100 centavos	68.6		1 Escudo = £0.0146	1½p
Antigua				E. Car. Dollar of 100 cents	4.80		1 E. Car. Dollar=£0.208	21p
Argentina	ISO	M	1887	Peso of 100 centavos	834		1 Pesa = £0.001 2	—
Australia	ISO	Imp/M		Aust. Dollar of 100 cents	2.147½		1 Aust. Dollar = £0.466	46½p
Austria	ISO	M	1867	Schilling of 100 groschen	61.95		1 Schilling = £0.016 1	1½p
Azores				Portugese Escudo of 100 centavos	68.60		1 Escudo = £0.014 6	1½p
Bahamas				B. Dollar of 100 cents	2.449		1 Dollar = £0.408	41p
Bahrain		Imp		Dinar of 1000 fils	1.142		1 Dinar = £0.874	87½p
Balearic Is.				Spanish Peseta of 100 centimos	157⅛		1 Peseta = £0.005 98	½p
Barbados				E. Car. Dollar of 100 cents	4.80		1 E. Car. Dollar = £0.208	21p
Belgium	ISO	M	1820	Belgian Franc of 100 centimes	119.20		1 Belgian Fr. = £0.008 4	1p
Bermuda				Bermudan Pound of 20 shillings of 240 pence	1		1 Bermudan Pound=£1.00	£1.00
Bhuton				Indian Rupee of 100 paise	18.04		1 Rupee = £0.055 4	5½p
Bolivia		M	1870	Bolivian Peso of 1,000 bolivianos	28⅛		1 Bolivian Peso=£0.035 5	3½p

Botswana		M	1870	S.A. Rand of 100 cents	1.716		1 S.A. Pound = £0.583	58p
Brazil	ISO	M	1862	New Cruzeiro of 1,000 old cruzeiros	7.67		1 Cruzeiro = £0.130	13p
Britain		Imp/M	1970	British Pound Sterling of 20 shillings or 240 pence	1.0		Pound Sterling	£1.00
Brit. Honduras		Imp		Brit. Hond. Dollar of 100 cents	4.0		1 Brit. Hond. Dollar = £0.25	25p
Brunei		Imp		Brunei Dollar of 100 cents		32.63*d.*	1 Brunei Dollar = £0.136	13½p
Bulgaria	ISO	M	1802	Lev of 100 statinki	2.81		1 Lev = £0.355	35½p
Burma		M	1920	Kyat of 100 pyas		21*d.*	1 Kyat = £0.087 5	8½p
Burundi		M		Burundi Franc	208.60		1 Burundi Fr. = £0.004 81	½p
Cambodia		M	1910	Riel of 100 sen	84		1 Riel = £0.011 9	1p
Cameroon				Cameroon Franc	590		1 Cam. Fr. = £0.001 695	—
Canada	ISO			Canadian Dollar of 100 cents	2.60⅝		1 Can. Dollar = £0.384	38½p
Canary Is.				Spanish Peseta	167⅛		1 Peseta = £0.005 98	½p
Cape Verde Is.				Escudo of 100 centavos	68.60		1 Escudo = £0.014 6	1½p
Ceylon	ISO	Imp		Rupee of 100 cents		16.84*d.*	1 Rupee = £0.070 2	7p
Chile	ISO	M	1865	Escudo of 100 condores	15.05	—	1 Escudo = £0.066 5	6½p
China		M	1929	Yuan of 100 fen and People's dollar	5.908 5		1 Yuan = £0.169	17p
Colombia	ISO	M	1854	Peso of 100 centavos	38¾		1 Peso = £0.025 8	2½p
Congo Republic		M	1910	Zaire of 100 makuta	1.2		1 Zaire = £0.834	83½p
Costa Rica		M	1912	Colon of 100 centimos	10.924		1 Colon = £0.091 5	9p
Cuba	ISO	M	1858	Peso of 100 centavos	2.40¼		1 Peso = £0.416	41½p
Cyprus	ISO	M	1961	Cyprus Pound of 1,000 mils	1.0		Pound Sterling	100p=£1
Czechoslovakia	ISO	M	1876	Koruna of 100 heller	17.28		1 Koruna = £0.057 8	6p
Dahomey				Franc	590		1 Franc = £0.001 7	—
Denmark	ISO	M	1910	Krone of 100 öre	17.89		1 Krone = £0.056	5½p
Djibuti				Djibuti Franc	510		1 D. Fr. = £0.001 96	—
Dominica				E. Caribbean Dollar	4.80		1 E. Car. Dollar=£0.208	21p
Dominican Rep.		M	1913	Dominican Peso of 100 centavos	2.40		1 D. Peso = £0.416	41½p

TABLE 10.2 (*cont.*)

State or Nation		Meast. system Date of adoption	Currency system	Conversion factor to sterling: 'To the pound'	Conversion factor to sterling: Value of unit in pence	Conversion factor into decimal currency	Nearest round figure in decimal currency
Ecuador		M 1856	Sucre of 100 centavos	43.20		1 Sucre = £0.023 2	2p
Egypt		M 1939	Egyptian Pound	1.04		1 £E = £0.96	96p
Eire	ISO	Imp/M	Pound Sterling of 20 shillings or 240 pence	1.00		Pound Sterling	
El Salvador		M 1886	Colon of 100 centavos	6.00		1 Colon = £0.167	16½p
Eritrea		M	Ethiopian Dollar of 100 cents	6.00		1 Eh. Dollar = £0.167	16½p
Ethiopia		M	Ethiopian Dollar of 100 cents	6.00		1 Eh. Dollar = £0.167	16½p
Falkland Is.			Pound Sterling of 20 shillings or 240 pence	1.00		Pound Sterling	
Faroe Is.			Krone	17.89		1 Krone = £0.056	5½p
Fiji Is.			Fiji Pound of 20 shillings or 240 pence	1.04⅞		1 Pound Fiji = £0.955	95½p
Finland	ISO	M 1892	Markka of 100 pennis	10.04⅝		1 Markka = £0.095	9½p
Formosa			New Taiwan Dollar	96.24		1 Taiwan Dollar = £0.010 4	1p
France	ISO	M 1794	Franc of 100 centimes	11.80		1 Fr. = £0.084 7	8½p
Gambia			Gambian Pound of 20 shillings or 240*d.*	1.00		Gambian £ = £ sterling	
Germany E.		M 1872	M.D.N. Mark of 100 pfennig (OstM)	10.01		1 Ostmark = £0.099 8	10p
Germany W.	ISO	M 1872	D.M. of 100 pfennig (Deutsche M)	10.01		1 Deutsche Mark = £0.099 8	10p
Ghana	ISO		New Cedi of 100 New Pesewa	2.45		1 New Cedi = £0.403	41p

Gibraltar				Pound of 20 shillings or 240*d*.	1.00	Pound Sterling	
Gilbert and Ellice Is.				Aust. Dollar of 100 cents	2.1471½	1 Aust. Dollar = £0.465	46½p
Greece	ISO	M	1922	Drachma of 100 lepton	71⅝	1 Drachma = £0.013 95	1½p
Greenland				Danish Krone of 100 öre	17.89	1 Krone = £0.056	5½p
Grenada				Trinidad and Tobago Dollar of 100 cents	4.80	1 T. & T. Dollar = £0.208	21p
Guadaloupe				Franc of 100 centimes (Fr.)	11.80	1 Fr. = £0.084 7	8½p
Guam				U.S. Dollar of 100 cents	2.40¼	1 US Dollar = £0.416	41½p
Guatemala		M	1912	Gold Quetial of 100 centavos	2.40¼	1 Quetial = £0.413	41p
Guinea				Guinea Franc of 100 centimes	586	1 Guinea Franc=£0.001 7	—
Guyana				Guyana Dollar of 100 cents	4.80	1 Guyana Dollar = £0.208	21p
Haiti		M	1921	Gourde of 100 centimes	12.01¼	1 Gourde = £0.083 1	8p
Honduras		M	1912	Lempira of 100 centivos	4.80½	1 Lempira = £0.208	21p
Hong Kong		Imp		Hong Kong Dollar of 100 cents	16.421875	1 H.K. Dollar = £0.060 6	6p
Hungary	ISO	M	1876	Forint of 100 filler	28.18	1 Forint = £0.035 5	3½p
Iceland		M	1907	Kröna of 100 aurar	137	1 Kröna = £0.007 3	1p
India (1966)	ISO	M	1920	Rupee of 100 paise	18.04	1 Rupee = £0.588	5½p
Indonesia	ISO	M	1930	Rupiah of 100 sen	624	1 Rupiah = £0.001 6	—
Iran	ISO	M	1933	Rial of 100 dinars	179.87¼	1 Rial = £0.005 5	½p
Iraq	ISO	M	1933	Iraq Dinar of 100 fils	0.857	1 Dinar = £1.17	£1.17
Israel (Palestine)	ISO	M	1928	Israeli Pound of 100 agorot	8.40	1 Israel Pound = £0.119	12p
Italy	ISO	M	1861	Lira of 100 centesimi	1498	1 Lira = £0.000 16	—
Ivory Coast				Franc C.F.A.	590	1 Franc CFA = £0.001 7	—
Jamaica				Jamaican Pound of 20*s*. or 240*d*.	1.0	Pound Sterling	—
Japan	ISO	M	1909	Yen	869	1 Yen = £0.001 15	—
Jordan		M	1933	Jordanian Dinar of 1,000 fils	0.858	Jordanian Dinar = £1.16	—
Kenya				Kenya Shilling of 100 cents	17.17⅜	1 Kenyan Shilling = £0.058	6p
Korea N.	ISO	M	1940	Won	6.17	1 Won = £0.162	16p
Korea S.	ISO	M	1940	Won of 100 jeon	612	1 Won = £0.001 6	—

TABLE 10.2 (*cont.*)

State or Nation	Meast. system Date of adoption	Currency system	Conversion factor to sterling		Conversion factor into decimal currency	Nearest round figure in decimal currency
			'To the pound'	Value of unit in pence		
Kuwait ISO		Kuwaiti Dinar of 1,000 fils	0.857 15		1 Kuwaiti Dinar = £1.17	£1.17
Laos	M 1910	Kip of 100 ats	680.8		1 Kip = £0.001 47	—
Lebanon ISO	M 1935	Lebanese Pound of 100 piastres	7.39		1 Lebanese Pound = £0.135	13½p
Lesotho		S.A. Rand of 100 cents	1.716		1 S.A. Rand = £0.583	58p
Liberia		Liberian Dollar of 100 cents	2.40¼		1 Dollar = £0.416	41½p
Libya	M 1920	Libyan Pound of 1,000 milliemes	0.855		1 Libyan Pound = £1.17	£1.17
Liechtenstein	M 1870	Swiss Franc of 100 centimes	10.39½		1 Sw. Fr. = £0.096	9½p
Luxembourg		Franc of 100 centimes	119.20		1 Lux. Fr. = £0.008 4	1p
Macau	M	Pataca of 100 avos	14.63		1 Pataca = £0.068 4	7p
Madagascar ISO		Franc Malagäche (FMG)	590		1 Mal. Fr. = £0.001 7	—
Madeira		Portugese Escudo of 100 centavos	68.60		1 Escudo = £0.014 6	1½p
Malagasy		Franc FMG	590		1 Fr. FMG = £0.001 7	—
Malawi		Pound Sterling 20*s*. or 240*d*.	1.00		Pound Sterling	
Malaysia	M	Malaysian Dollar of 100 cents	32.63		1 Malay Dollar = £0.030 6	3p
Maldive Is.	M	Mal. Rupee	11.429		1 Mal. Rupee = £0.087 5	9p
Mali Republic		Mali Franc	1.176		1 Mali Fr. = £0.000 85	—
Malta and Gozo ISO	M 1921	Maltese Pound (Sterling)	1.00		Pound Sterling	
Martinique		Franc of 100 centimes	11.80		1 Mart Fr. = £0.08 47	8½p
Mauritania		Franc (CFA)	590		1 Fr. CFA = £0.001 7	—
Mauritius	M 1878	Rupee of 100 cents	13.30		1 M. Rupee = £0.007 5	1p
Mexico ISO	M 1928	Peso of 100 centavos	29⅞		1 Peso = £0.033 5	3p

Miguelon				Franc (CFA)	590	1 Fr. CFA = £0.001 7	—
Monaco		M	1850	French Franc	11.80	1 French Fr. = £0.085	$8\frac{1}{2}$p
Mongolia				Tugrik	9.60	1 Tugrik = £0.104	10p
Montserrat				E. Caribbean Dollar	4.80	1 E. Car. Dollar=£0.208	21p
Morocco	ISO	M	1920	Dirham of 100 francs	$12.12\frac{1}{2}$	1 Dirham = £0.008 25	8p
Mozambique				Escudo of 100 centavos	68.60	1 Escudo = £0.014 6	$1\frac{1}{2}$p
Muscat and Oman		Own system		Gulf Rupee of 100 naya paisa	13.285	1 Gulf Rupee = £0.075 5	$7\frac{1}{2}$p
Nauru Is.				Australian Dollar	2·147 $1\frac{1}{2}$	1 Aust. Dollar = £0.465	$46\frac{1}{2}$p
Nepal				Nepalese Rupee	24.301 3	1 Nep. Rupee = £0.041 2	4p
Netherlands	ISO	M	1820	Guilder (Florin) of 100 cents	$8.64\frac{1}{2}$	1 Guilder = £0.116	$11\frac{1}{2}$p
Netherlands Guiana				Surinam Guilder	4.55	1 Sur. Guilder = £0.22	22p
Netherlands W. Indies				Antillian Guilder	4.55	1 Ant. Guilder = £0.22	22p
New Guinea		M	1920	Australian Dollar	$2.1471\frac{1}{2}$	1 Aust. Dollar = £0.465	$46\frac{1}{2}$p
New Hebrides				CFP Franc	214.27	1 Fr. CFP = £0.009 67	$\frac{1}{2}$p
New Zealand	ISO	Imp/M		N.Z. Dollar of 100 cents	2.142 $9\frac{1}{2}$	1 N.Z. Dollar = £0.465	$46\frac{1}{2}$p
Nicaragua		M	1912	Cordoba of 100 centavos	16.80	1 Cordoba = £0.059 5	6p
Niger Rep.				Franc CFA	590	1 Fr. CFA = £0.001 69	—
Nigeria				Nigerian Pound of 20 shillings/240*d.*	0.857 12	1 Nig. £ = £1.17	£1.17
North Borneo		Imp					
Norway	ISO	M	1882	Krone of 100 ore	17.13	1 Krone = £0.058 3	6p
Pakistan	ISO	Imp/M		Rupee of 100 paisa	21	1 Rupee = £0.047 6	5p
Panama		M	1910	Balboa (= U.S. Dollar) of 100 cents	$2.40\frac{1}{4}$	1 Balboa = £0.416	$41\frac{1}{2}$p
Panama Canal Zone				U.S. Dollar of 100 cents	$2.40\frac{1}{4}$	1 U.S. Dollar = £0.416	$41\frac{1}{2}$p.
Papua				Australian Dollar	2.147 $1\frac{1}{2}$	1 Aust. Dollar = £0.465	$46\frac{1}{2}$p
Paraguay	ISO	M	1901	Guarani of 100 centimas	297	1 Guarani = £0.003 37	—
Peru	ISO	M	1869	Sol of 100 centavos	$98\frac{1}{2}$	1 Sol. = £0.010 2	1p

TABLE 10.2 (*cont.*)

State or Nation		Meast. system Date of adoption	Currency system	Conversion factor to sterling: 'To the pound'	Conversion factor to sterling: Value of unit in pence	Conversion factor into decimal currency	Nearest round figure in decimal currency
Philippines		M 1860	Philippine Peso of 100 centavos	9.37		1 Phil. Peso = £0.107	10½p
Pitcairn Island			Pound Sterling and N.Z. Dollar				
Poland	ISO	M 1910	Zloty of 100 groszy	9.6		1 Zloty = £0.104	10½p
Port Guinea			Escudo of 100 centavos	68.60		1 Escudo = £0.014 6	1½p
Port Timor			Escudo of 100 centavos	68.60		1 Escudo = £0.014 6	1½p
Portugal	ISO	M 1872	Escudo of 100 centavos	68.60		1 Escudo = £0.014 6	1½p
Principe Is.			Guinea Escudo	68.60		1 Escudo = £0.014 6	1½p
Puerto Rico		M 1840	U.S. Dollar	2.40¼		1 U.S. Dollar = £0.416	41½p
Qatar			Qatar Dubai Riyal of 100 dirhams	11.54		1 Q.D. Riyal = £0.086 6	8½p
Reunion, Île de la			Franc (CFA)	590		1 Fr. CFA = £0.001 7	—
Rhodesia			Pound of 20*s*. or 240*d*.	n. a.		n. a.	
Rumania		M 1884	Leu of 100 bani	19.4		1 Leu = £0.069 5	7p
Rwanda			Rwanda Franc	240		1 Rwanda Fr. = £0.004 16	½p
St. Christopher			E. Caribbean Dollar	4.80		1 E. Car. Dollar = £0.208	21p
St. Helena			Pound Sterling	1.00		Pound Sterling	
St. Lucia			E. Caribbean Dollar	4.80		1 E. Car. Dollar = £0.208	21p
St. Pierre			Franc (CFA)	590		1 Fr. CFA = £0.001 7	—
St. Vincent			E. Caribbean Dollar	4.80		1 E. Car. Dollar = £0.208	21p
St. Vincent			U.S. Dollar	2.40¼		1 U.S. Dollar = £0.416	42½p
San Marino		M 1900	Italian Lira	1,498		1 Lira = £0.000 66	—

Sao Tome				Guinea Escudo	68.60		1 Guinea Escudo = £0.014 6	$1\frac{1}{2}$p
Sarawak		Imp						
Saudi Arabia		M		Ryal	n. a.		n. a.	
Seychelles		M	1880	Seychelle Rupee of 100 cents	13.33		1 Seychelle Rupee = £0.075	
Sierra Leone				Leone of 100 cents	2.00		1 Leone = £0.50	50p
Singapore	ISO			Singapore Dollar	32.63		1 Singapore Dollar = £0.030 6	3p
Somalia		M		Somali Shilling of 100 cents	17.141 5		1 Som. Shill = £0.058 3	6p
South Africa	ISO	Imp/M		Rand of 100 cents	1.716		1 S.A. Rand = £0.583	58p
South Arabian Protectorates				South Arabian Dinar of 1,000 fils	1.00		1 Dinar = £1.00	£1.00
S.W. African Territories				S.A. Rand	1.716		1 S.A. Rand = £0.583	58p
Southern Yemen				South Arabian Dinar	1.00		1 Dinar = £1.00	£1.00
Spain	ISO	M	1859	Peseta of 100 centimas	$167\frac{1}{8}$		1 Peseta = £0.005 98	$\frac{1}{2}$p
Spanish Guinea				Peseta	$167\frac{1}{8}$		1 Peseta = £0.005 98	$\frac{1}{2}$p
Spanish Morocco				Peseta	$167\frac{1}{8}$		1 Peseta = £0.005 98	$\frac{1}{2}$p
Spanish Sahara		M		Peseta	$167\frac{1}{8}$		1 Peseta = £0.005 98	$\frac{1}{2}$p
Sudan Rep.		M	1950	Sudanese Pound of 100 piastres	0.835 8		1 Sudan £ = £1.195	£$1.19\frac{1}{2}$
Surimam		M	1910	Gulden or Florin of 100 cents				
Swaziland		Imp		S.A. Rand	1.716		1 Rand = £0.583	58p
Sweden	ISO	M	1878	Kronor of 100 öre	$12.42\frac{1}{2}$		1 Kronor = £0.080 5	8p
Switzerland	ISO	M	1875	Franc of 100 centimes	$10.39\frac{1}{2}$		1 Sw. Fr. = £0.096 3	$9\frac{1}{2}$p
Syria	ISO	M	1955	Syrian Pound of 100 piastres	$10\frac{1}{4}$		1 Syria £ = £0.097 5	10p
Tangier		M						
Tanzania		Imp		Tan Shilling of 100 cents	17.109 3		1 Tan Shill. = £0.058 5	6p
Thailand	ISO	M	1923	Tical of 100 stangs (Tical = Baht)	$49.57\frac{1}{2}$		1 Tical = £0.020 2	2p
Tibet				Chinese People's Dollar	5.908 5		1 Chinese Dollar = £0.169	17p

TABLE 10.2 (*cont.*)

State or Nation		Meast. system Date of adoption	Currency system	Conversion factor to sterling: 'To the pound'	Conversion factor to sterling: Value of unit in pence	Conversion factor into decimal currency	Nearest round figure in decimal currency
Togo Republic			Franc (CFA)	590		1 Fr. CFA = £0.001 7	—
Tonga Is.			Pa'anga of 100 senti	2.143		1 Pa'anga = £0.466	46½p
Trinidad and Tobago			T. and T. Dollar of 100 cents	4.80		1 T. and T. Dollar = £0.208	21p
Trucial Sheikdoms			Qatar Dubai Riyal	n. a.		n. a.	
Tunisia		M 1895	Tunisian Dinar of 1000 milliemes	1.261 8½		1 Tun. Dinar = £0.792	79p
Turkey	ISO	M 1934	T. Pound of 100 piastres	21.6		1 T. Pound = £0.046 3	4½p
Uganda			U. Shilling of 100 cents	17.17⅜		1 Ug. Shill = £0.058 2	6p
U.S.S.R.	ISO	M 1927	Rouble of 100 copeks	2.16		1 Rouble = £0.462	46p
United Arab Republic	ISO		Pound of 100 piastres	1.04		1 UAR Pound = £0.96	96p
U.S.A.	ISO	Imp/M	Dollar of 100 cents	2.40¼		1 U.S. Dollar = £0.415	41½p
Upper Volta Republic			Franc (CFA)	590		1 Fr. CFA = £0.001 7	—
Uruguay		M 1894	Peso of 100 centesimos	476½		1 Peso = £0.002 1	—
Vatican			Italian Lira	1,498		1 Lira = £0.000 66	—
Venezuela	ISO	M 1914	Bolivar of 100 centimos	10.70		1 Bolivar = £0.093 4	9p
Vietnam North		M 1910	Dong of 100 xu	8.83		1 N. Dong = £0.113	11p
Vietnam South		M 1910	Dong of 100 cents	192		1 S. Dong = £0.005 2	½p
Virgin Island			U.S. Dollar	2.40¼		1 U.S. Dollar = £0.416	41½p
West Indies, The			W.I. Dollar of 100 cents	4.80		1 W.I. Dollar = £0.208	21p

Western New Guinea				New Guinea Guilder	8.64½		1 N.G. Guilder = £0.116	11½p
Western Samoa				Samoan Thaler				
Yemen				Riyal	n. a.		n. a.	
Yugoslavia	ISO	M	1883	Dinar of 100 paras	30		1 Dinar = £0.332	33p
Zambia				Zambian Pound of 20*s*. or 240*d*.	1.714 25		1 £Zambian = £0.583	58p
Zanzibar and Pemba		Imp						
Icaiti (Central Am.) (Costa Rica, El Salvador, Nicaragua, Panama)	ISO	M						
Br. Solomon Is.				Australian Dollar of 100 cents	2.147 1½		1 Aust. Dollar = £0.466	46½p
Br. Virgin Is.				U.S. Dollar of 100 cents	2.40¼		1 U.S. Dollar = £0.416	41½p

n. a. = not available

CHAPTER 11

Metrication and Modular Construction

11.1. Modular Sizes

The introduction of industrialized building methods, together with the growing use of new materials and structural sections, has created an urgent need for coordination of the dimensions of building components.

The method being adopted is that of modular sizes. The modules recommended for general use are, as choices,

first	3n decimetres,
second	1n decimetre,
third	0.5n decimetre up to 3 dm,
fourth	0.25n decimetre up to 3 dm.

These values are according to BS 4011 : 1966, *Basic Sizes*. Prefabricated components such as door and window frames, wall panels, floor slabs and the like should be manufactured to the modular sizes and like components in different materials should be to the same module.

Where a number of components assemble, the overall size of the assembly should be a modular size.

The construction industry has set itself the target date of 1972 for the virtual completion of the changeover to metric modular construction. This is a tremendous undertaking because the insistence on modular sizes, to round numbers of millimetres, makes pseudo-metrication impossible. The actual hardware will have to be changed in order to comply.

11.2. The Metric Brick

The Metric Brick is a problem which has produced a variety of conflicting proposals. The Brick Development Association (BDA) have stated their case as follows:

Only recently has a single national standard size been settled by the issue of a British Standard. The Current Standard, BS 3921: 1965, incorporating Amendment 1, states that the standard brick shall be

$$8\tfrac{5}{8} \text{ in. by } 4\tfrac{1}{8} \text{ in. by } 2\tfrac{5}{8} \text{ in.}$$

With joints this becomes

$$9 \text{ in. by } 4\tfrac{1}{2} \text{ in. by } 3 \text{ in. (ratio } 6:3:2).$$

The BDA propose that the metric brick shall be

215 mm by 102.5 mm by 65 mm.

With joints this becomes

225 mm by 112 mm by 75 mm (ratio 6 : 3 : 2).

The BDA wish to maintain the present size by this 'pseudo-metrication' in order to make extension and repair work possible. They also wish to maintain the ratio: twice width plus joint equals length plus joint.

The Modular Society point out that the BDA proposal fails to meet the requirements of BS 4011 at every stage except one, thus:

Vertically		BS 4011 preference
12 in.	3.00 dm	1st
9 in.	2.25 dm	4th
6 in.	1.50 dm	3rd
3 in.	0.75 dm	4th

which is excellent.

But, horizontally:

Feet	Inches	Decimetres	BS 4011 preference
3	0	9.00	1st
2	7½	7.875	X
2	3	6.750	4th
1	10½	5.625	X
1	6	4.500	3rd
1	1½	3.375	X
1	0	3.00	limit
0	9	2.25	4th
0	4½	1.125	X

This, say the Modular Society, is not good enough.

The Modular Society's own proposals are as follows:

Sizes in decimetres:
3 by 1 by 1
2 by 1 by 1
2 by 1 by 0.75
2 by 2 by 1

The first will produce the most economical brickwork.

The second is the ordinary modular brick, available in the U.S.A. and Canada as 8 in. by 4 in. by 4 in.

11.3. Changeover Programme for the Construction Industry

The construction industry has elected to carry out the bulk of its changeover by the end of 1972. This should mean that, in general, corresponding events in the construction industry programme should be at least two years ahead of those in the manufacturing industry.

The British Standards Institution has acted as the focus for the changeover and the programme has been prepared by the Metric Panel (Construction Industry). The programme has been approved by the Building Divisional Council and published under the authority of the General Council on 21st February 1967 as document Number PD 6030.

One of the most fundamental documents required will be a set of key dimensional recommendations to link up the bare statement of modular sizes given in BS 4011, *Basic Sizes for Building Components*, with the detailed dimensions of a range of components.

11.4. The Building Regulations

Another basic requirement before architects can design and draw in metric units is the conversion of the Building Regulations. The Building Regulations is a statutory document and, as such, requires action by Parliament to change it. When the conversion of its terms is complete the metric system will indeed have become the legal system of measurement in this country, since Parliament will have to stipulate an instant of time at which the new version becomes law, and from that instant, in so far as Building Regulations are concerned, there will be no other legal system. The fundamental nature of this change means that it cannot take place until the standardization exercise is virtually complete, which, in turn, means that it cannot occur until fairly late in the changeover period.

The authoritative document describing the technicalities and effects of the changeover as it affects the construction industry is the *Metric Handbook*, published by the *Architect's Journal*. (*AJ Metric Handbook*, price 15*s*., The Architectural Press Ltd., 9–13 Queen Anne's Gate, London S.W.1.)

The Royal Institution of Chartered Surveyors has published a *Metric Guide and a set of Metric Conversion Tables* to guide their members through the change. (Royal Institution of Chartered Surveyors, 12 Great George Street, Parliament Square, London, S.W.1.)

Architects and surveyors have to refer to a number of statutory documents in the course of their work and serious confusion will occur if changes in the building regulations and British Standards are not matched by simultaneous changes in such things as Ballast Regulations, Schedules 5 and 7 of the Weights and Measures Act, Industrial Development Certificates, Town Planning Acts, Factory Acts and so on.

CHAPTER 12

Metrication in the Electrical Industry

12.1. The Professional Electrical Engineer

Rationalized MKSA units, also known as the Giorgi System, were introduced by the International Electrotechnical Commission in 1935 and have been in common use for over twenty years. The basic textbooks of electrical physics and technology have long been available in MKS editions.

The only changes the professional electrical engineer will need to make are the adoption of the SI preferred multiples, the adoption of the name 'tesla' for the unit of flux density, and the name 'hertz' for the unit of frequency.

12.2. The Interface with Mechanical Engineering

Mechanical engineers and those electrical engineers primarily concerned with rotating machines will have a number of changes to get used to. The horsepower will be replaced by the kilowatt and the revolution per minute by the revolution per second. These changes will bring about great simplifications but doubtless there will be some initial mental discomfort.

12.3. Electrical Manufacturing Industry

Metric standards in electrical manufacturing industry will be no novelty. The International Electrotechnical Commission was formed in 1906 and IEC Recommendations have, in a great many cases, formed the basis for British Standards for many years.

BS 3979: 1966, *Dimensions of Electric Motors* (*Metric Series*), to quote but one example, agrees with IEC Recommendation 72: 1959. The U.K. home market has accepted metric shafts, keyways, shaft heights and fixing dimensions without any detectable difficulty.

Copper and aluminium conductor sizes tend to be based on convenient current-carrying capacities and hence the physical dimensions are not, usually, 'round' numbers. The change to metric dimensions is merely a case of direct conversion of existing imperial sizes. British Insulated Callender's Cables Limited have published a booklet, *Electrical Conductors, International Standard Sizes*. The difference between the old cable standards and the new is that, generally, less insulation is used on the new cables.

Plastic mouldings and insulating materials can, of course, be made in any reasonable physical dimensions and no difficulty arises from specifying dimensions in millimetres. The plastics industry is, in any case, accustomed to working in metric units.

12.4. Illumination Engineering

The sixth edition of the *Code for the Interior Lighting of Buildings*, published by the Illuminating Engineering Society (IES), uses the SI unit of illumination, the lux, to express its standards. The standard general office lighting level, for example, is quoted as 400 lux (1 lumen/foot^2 = 1 foot-candle = 10.763 9 lux).

Metric standards exist, and are in use, for such items as lamp-holders and lamp caps, lighting fittings and tubular fluorescent lamps.

BS 3820: 1964, *Electric Lighting Fittings*, for example, agrees with IEC 162: 1965, *Lighting Fittings for Tubular Fluorescent Lamps*. The physical dimensions and photometric requirements in these two documents agree.

12.5. Electrical Installations

Conduit and conduit threads to Imperial Standards (BS 31: 1940 with Amendments CF 9846 and PD 5663) are used both in

Britain and on the Continent. It seems unlikely that any rapid changeover to metric sizes and threads will occur.

A standard on metric conduit and conduit threads does exist. It is issued by the International Commission on Rules for the Approval of Electrical Equipment and is designated CEE 23: 1962, *Specification for Steel Conduits and Fittings for Electrical Installations*. The size range covered is 16 mm to 60 mm diameter in both light gauge plain and heavy gauge screwed series. The thread is ISO 60° 1.5 mm pitch.

12.6. Electronic Manufacturing Industry

The situation in the electronic industry is complicated by the fact that the United States of America is overwhelmingly powerful in terms of financial and technical control, and is the source of much of the technical innovation.

The printed-circuit technique has grown into a large-scale industry based on inch standards. The pitch of the gold-plated connector strips along the edge of printed circuit boards is in multiples of one-tenth of an inch. This standard is used on the Continent as well as in Britain and America.

Multi-pin, screened, and coaxial connectors for electronic equipment are very largely based on American design and technique. The U.S. Military Specifications and Standards are familiar to most workers in this field. The dimensional modules are, therefore, inch-based. Many European manufacturers, in order to comply with U.S. Mil-Spec requirements, offer inch-based designs. It seems unlikely that, in the near future, any fundamental change in this situation will occur. Announcement of an American Metrication Programme would, of course, reopen the whole question.

It is also interesting to note that the British Post Office 19-inch rack is accepted as a standard on the Continent. This modular construction system utilizes a standard width (19 inches) and modular heights, which are multiples of $1\frac{3}{4}$ inch. This is another case where the acceptance of a standard, and the existence of large quantities of hardware built to that standard, make any radical change unlikely in the forseeable future.

12.7. Electrical Generating Plant

The Central Electricity Generating Board have informed the Ministry of Power, through the Electricity Council, of their support towards the adoption and subsequent implementation of the change to SI by 1975 so far as is economic and practicable.

The CEGB have set up a Metric Steering Committee to work out policy for new construction work. A 'notional plan' has been published outlining desirable time scales for the introduction of metric practice.

All power-stations and associated projects commissioned in and after 1975 should be in an advanced state of metrication but, in view of the number of contractors and sub-contractors involved, very careful planning and much mixed design will be required.

The life of large capital projects of the power-station type is of the order of thirty to forty years and there will, therefore, be a demand for imperial spare parts, tools, measuring instruments and instruction manuals throughout the period 1975 to 2015.

The first step towards metrication of generating plant has already been taken: from 1st January 1967 CEGB have expressed a preference for ISO Metric fasteners. From 1st January 1970 metric units will be used throughout in new designs.

CHAPTER 13

Metrication in Mechanical Engineering

13.1. The Professional Mechanical Engineer

The disappearance of familiar units such as the horsepower and the pound-per-square-inch will mean a considerable loss of confidence in calculated values at first. The amount of rule-of-thumb working and memorizing of safe limiting values only becomes apparent when the familiar units and quantities are lost. Some comfort may be gleaned from easy conversions, such as the fact that 1 U.K. ton = 1 tonne within 1.6% error, and 1 tonf/in² = 1.5 hectobar within 3% error. It is important to realize, however, that the SI units are the ones which we shall all be using for the rest of our working lives and that conversion back into Imperial is likely to delay the familiarization process.

13.2. Material Testing

The fundamental prerequisite for an engineering design is a knowledge of the strength and properties of the materials available. A survey of the literature and standards concerned with testing will therefore indicate which units will gain common acceptance.

The metric version of BS 18, the tensile testing of metals, uses the hectobar as the stress unit. The corresponding ISO Recommendation R82, *Tensile Testing of Steel*, still, in the February 1959 edition, uses the old metric technical stress unit, the kgf/mm².

The load applied to a specimen must, of course, be expressed in units of force and the appropriate unit is the newton. It is convenient that 1 ton-force is very nearly equal to 10 kilonewtons

(1 tonf = 9.964 kN). This enables one to read a ton-force scale in kilonewtons merely by mentally adding a nought.

Impact testing of the Izod type requires a unit of energy. The familiar foot-pound will give way to the joule (1 ft-lbf = 1.356 J). Fracture toughness testing units have been the subject of much debate but a BISRA announcement on 1st August 1968 favours the hectobar $\sqrt{}$centimetre for the measurement of plane strain fracture toughness, in place of the present lbf/in$^2\sqrt{\text{in}}$. The BISRA announcement wisely adds that the pure SI unit should be shown in a note on each page of text thus:

$$1 \text{ hbar}\sqrt{\text{cm}} = 1 \text{ MN}^{-\frac{3}{2}}.$$

From the basis given above other specialized units can be derived, the examples quoted cover the vast majority of cases.

13.3. Preferred Sizes

The real incentive for metrication is compatibility and variety reduction: compatibility with the units, standards and practices of the majority of mankind, and the elimination of two lines of production, 'home' and 'export'.

In addition to the elimination of Imperial products the introduction of ISO/SI practice reduces variety by establishing rationalized international standards to replace the previous multiplicity of national standards.

Realistic studies of stockholding policies and practices in several industries have led to the conclusion that the cost of the metrication exercise can in many cases be more than covered by the economies resulting from variety reduction.

Initially it is necessary to prepare stockholding charts showing, on a large scale, categories of stock, and on a small scale, individual items. The usual storeman's technique of using a colour code to indicate minimum quantities is useful in this context. When a complete set of charts is to hand it is essential to survey the items to identify 'dead' stock, there are very few organizations without some lines which turn over at an unreasonably slow rate. Items ordered as specials for a particular job tend to persist in stock lists long after their retention can be justified. As a first step any item which has not turned over in three years can be deleted from the charts. As a next exercise the low-usage items can be surveyed and their occasional users can be invited to make out a case for their retention. Failing the presentation of a good case the item can be deleted.

When the preliminary combing through is complete the charts must be translated into metric terms, the nearest equivalents in metric preferred sizes being set down. This stage will, in itself, usually reduce the total number of items remaining in the charts. To quote some examples from the world of fasteners:

ISO Metric should be used, and wherever possible only ISO Metric Coarse threads should be regarded as normal stock. ISO-Inch, or Unified threads as they were previously known, should only be considered if the existing stock range is predominantly Unified.

BSW, BSF and BA should be phased out as quickly as possible, they are obsolete.

Specify not merely preferred lengths and diameters but, wherever possible, your supplier's stock lengths and diameters.

Restrict the range of tensile strengths held in stock, enquire what objections there would be to stocking only high-tensile bolts, eliminating black bolts. This does not automatically put up costs and is worth investigating.

The nine sizes from 0BA to 8BA have been replaced by five metric sizes.

M12 is regarded as a replacement for both $\frac{7}{16}$ in. and $\frac{1}{2}$ in.

Similar guidelines can be laid down for other items such as pipes, flats and sections. The main requirement is firmness in resisting dubious claims.

The relevant standard for range selection is BS 4318: 1968.

13.4. Gears

The ISO Gear System is a modular system. In the past most British engineers have worked in terms of diametral pitch. Diametral pitch is defined as the ratio of the number of teeth to the number of inches of pitch diameter, that is:

Diametral pitch = no. of gear teeth per inch of pitch diameter.

The concept used in the metric world is that of the module, which is defined as the ratio of pitch diameter to the number of teeth. It is understood that the pitch diameter is in millimetres.

Module = pitch dia. (mm) : no. of teeth.

This change naturally results in the loss of some familiar figures but the new form is quickly assimilated.

Consider a diametral pitch of 10:

$$10 = \frac{\text{no. of teeth}}{\text{pitch dia. in inches}}.$$

20-tooth wheel : diameter 2 inches.
Now: 2 inches is roughly 50 millimetres.

$$\text{Module} = \frac{\text{pitch dia. in mm}}{\text{no. of teeth}}.$$

$$= \frac{50}{20} \simeq 2.5.$$

The rough equivalent of a diametral pitch of 10 is, therefore, a module of 2.5.

A diametral pitch of 1:

$$1 = \frac{\text{no. of teeth}}{\text{pitch dia. in inches}}.$$

20-tooth wheel : diameter 20 inches.
Now: 20 inches is roughly 500 millimetres.

$$\text{Module} = \frac{500}{20} = 25.$$

13.5. Pipes, Tubes and Fittings

Pipe and pipe threads provide an example of pseudo-metrication, in that the international standards are inch-based. ISO/R7,

Pipe Threads for Gas List Tubes and Screwed Fittings where Pressure-tight Joints are made on the Threads ($\frac{1}{8}$ *in. to* 6 *in.*), is based on BS 21. The current edition of BS 21, issued in 1957, retains the inch values as the standard figures and quotes corresponding rounded metric figures.

BS 10, the standard on pipe-flanges, is currently (mid-1968) undergoing revision and it is obviously extremely unwise to embark on pipe work design incorporating ISO Metric fasteners until the metric version of the flange standard is to hand.

Pumps, which are associated with pipework, will have their power input requirements stated in watts or kilowatts instead of the old horsepower. Pressures will be in the newton per square metre and its multiples, heads will be in metres. The discharge of a pump will be stated in cubic metres per second.

13.6. Machinery Details

A survey of this large subject is not possible and an arbitrarily selected example of the kind of snag which may arise will be considered. A great many metal cutting tools are located and driven by means of a taper, the familiar standard in the Imperial world is the Morse taper.

Now, 0 to 6 Morse tapers are very similar to 5% metric tapers and are commonly tabulated together. ISO Recommendation No. 296 covers self-holding tapers for tool shanks. Those designated 1 to 6 Morse are not all identical with those listed under the same heading in BS 1660, Part 1, *Self Holding Tapers*. However, BS Handbook 18 1966, *Metric Standards for Engineering:* tabulates a set of metric taper dimensions which ensure interchangeability with components manufactured to the inch dimensions given in BS 1660, Part 1. The taper dimensions do not, unfortunately, complete the story; there is often a screw thread associated with a taper, provided for use with a draw bar. The recommended thread for this purpose is ISO Metric Coarse. There are, however, many components in existence having tapers designated as Morse and having Unified Coarse threads. Thus, even though the taper may be interchangeable the drawbar is not. The component must be identified by reference to the letter M signifying an ISO Metric Coarse thread or UNC signifying a Unified Coarse thread.

13.7. Measuring Instruments

The first fact to grasp about metric measurements in mechanical engineering is that the millimetre is a better unit than the inch. Because a millimetre is only one twenty-fifth of an inch, or forty thousandths of an inch, it is possible to express dimensions to the same precision with one less significant figure. The adoption of metric measurement can, therefore, often lead to greater economy of expression.

The introduction of a new basic unit of length is a wonderful opportunity to take a critical look at the accuracy of commonly used workshop measuring techniques. Practical tests show that micrometers are usually read low, that is to say that too much compression is applied to the screw. Ratchet devices avoid massive overtightening but even a ratchet, if overwound rapidly, can apply too much force. The book *Engineering Tolerances* by H. G. Conway, published by Pitman, contains an Appendix which summarizes the results of micrometer reading tests. The conclusion drawn from the results is that a skilled operator can read to ± 0.000 2 in. (± 5μm). This leads on to the conclusion that any component whose tolerance is less than 0.002 in. (50 μm) cannot realistically be measured with a micrometer since the reading error alone, apart from errors in the micrometer itself, amounts to at least 20% of the tolerance.

To the sources of error resulting from the handling and reading of the micrometer must be added sources of error within the instrument itself. External micrometers to BS 870 have a screw pitch of 0.5 mm. The range of error of traverse of the screw, for micrometers up to 300 mm measuring range, is 3 μm. The zero reading may be expected to be within 3 μm. Thus a total of 9 μm can realistically be added to the handling and reading errors to bring the total to 19 μm, say 20 μm. (This is equivalent to 0.000 8 in.)

Vernier gauges may be considered along similar lines. Vernier callipers to BS 887 are permitted a maximum error of ± 20 μm. Thus tolerances of less than 400 μm (0.015 mm) are in a class where methods other than the vernier calliper ought to be considered for final checking.

Under workshop conditions, where dirt, wear and rough handling all adversely affect the accuracy of measurement, the above figures are not unduly pessimistic.

It is probably true to say that many draughtsmen apply excessively severe tolerances largely through lack of a systematic approach to measuring and gauging practice and the economics of production. It is highly desirable that blind conversion of established tolerances should be avoided and that metrication should be made an occasion for critical reviews of design and inspection policy.

CHAPTER 14

Education and Training

14.1. The Three Phases of Education

For the purposes of this exercise we may consider education as taking place in three phases.

1. General, full-time primary and secondary education.
2. Professional, vocational or craft (further) education.
3. Specialized training undertaken while in employment.

1. General Education

The Schools Council is the body bearing overall responsibility for the curricula of primary and secondary schools in Britain. The Schools Council issues advice to various GCE and CSE examining bodies.

The great problem facing the schools council is that there exists a discrepancy between the announced intentions of manufacturing industry on the one hand and the absence of any announced intentions by the retail sector on the other.

It has, of course, always been recognized that there must be a period of bilinguality in both currency and weights and measures. All authorities are, however, agreed that the period should be as short as possible. It is therefore important to determine the earliest date by which wholly decimal-metric-trained school-leavers can be released onto the labour market without the risk of their encountering remaining pockets of Imperial practice.

The elimination of Imperial practice from school curricula will, so informed estimates suggest, result in a gain of approximately one academic year.

2. *Further Education*

The situation in further education is considerably simpler than that in general education. The appropriate time at which to revise the syllabus for a particular technical course is easily determined by examining the detailed metrication programme for the associated sector of industry. The important fact to remember during the changeover period is that although students leaving before the completion of metrication in their particular industry will still require to be conversant with Imperial units, they will be using metric units for the remainder of their careers. It is therefore vital that a rapidly growing emphasis be placed on metric units.

It is also important to ensure that teaching staff, particularly older staff, recognize that the primary aim is to train students in the use of metric units, not to familiarize them with conversion from one system to another.

The SI is to be the language of examination papers in technical subjects at the latest by mid-1970, in many cases by mid-1969.

It is to be hoped that twelve months notice will be given to teaching staff before the use of metric units becomes the only permissible practice in examinations.

3. *Specialized Training*

It must be recognized that it is possible to overemphasize formal training. It is, of course, vitally necessary to make people aware of the fact of metrication, and of the date upon which the practice of their craft is to be conducted in metric units. Some authorities, however, have tended to advise the preparation of rather elaborate training schemes for numerous categories of staff.

Workshop staff of a reasonable level of skill perform many calculations in the course of a day's work, and it is all too easy to underestimate the capability of a particular type of worker.

The tool-room technician, for example, performs elaborate

decimal arithmetic calculations in setting out angles with the aid of slip gauges and sine bar. The change in the size of the basic unit of length is a simplification to such a man. It seems unlikely that any useful purpose would be served by lengthy theoretical talks to such staff.

What is necessary is to explain to staff why the change is being made, emphasizing the overwhelming world swing to metric usage and the advantages of conformity with the majority of mankind.

Beyond an introductory explanation of the basic scales on the metric versions of the measuring devices to be used there seems little else that one can tell the skilled man. Practice and familiarity with metric units will come quickly once the Imperial units disappear from drawings and Imperial measuring instruments are withdrawn from use.

Professional and design staff need only be told which units they are to use. This is probably best done by brief lectures and the issue of data sheets or booklets of conversion factors. The main emphasis should be on the timing of the start of true metric design, in metric modules, after which conversion back to Imperial quantities becomes a chore, to be dropped as soon as possible. No real assistance can be given to the man who has to learn to think in terms of metric units. The process is one of learning, not of teaching.

It is possible to frighten people by too great an air of mystery induced by solemn gatherings and long lectures. Units of measurement are a means to an end, if one can learn and use a formidable array of units such as the Imperial system then one can certainly quickly acquire facility in the vastly simpler SI.

14.2. Textbooks

The task facing publishers of technical books is, in the absence of direct financial aid, a formidable one. Mechanical and civil engineering, physics and building are the broad categories requiring complete re-equipment with SI textbooks.

All reputable electrical engineering books in print now have MKSA, that is basically SI, editions. Only a small amount of editing, to introduce such units as the hertz and the tesla, is involved.

Owners of costly private libraries of technical books are faced with the realization that their volumes are outdated, although of course this fate overtakes all but a few classical texts in any case. It is of course possible to work through a book writing in SI conversions for all stated quantities but this process is not the ideal way to familiarize oneself with SI.

14.3. Training Aids

Numerous visual aids can be made or purchased by those responsible for staff training. Associability charts are poster-style sheets which present, pictorially, some familiar object from the industrial or social environment of the viewer, and which present the key dimension of the object in SI units.

Examples of suitable subjects for associability charts would be:

A telephone dial with the diameter shown in millimetres.

A 0–1-inch (0–25 mm) micrometer with the measuring range shown in millimetres.

A pedal cycle with the wheel diameters shown in millimetres. Many such common objects can be used both in pictorial form and in reality. The labelling of doorway sizes, for instance, can help to familiarize staff with the metre.

Most authorities stress the desirability of avoiding, as far as possible, the return of staff, who have been retrained in SI units, to Imperial work. Ideally this situation should be avoided but in practice it is almost bound to arise. One thinks of the necessity, at some time in the future, of making spares for Imperial equipment. The best way out of such a situation will probably be for the drawing office to convert the old Imperial drawing to metric units so as to avoid reminding the workshops of the old units.

Apprentices in training workshops should certainly not be taught conversion from one system to another until near the end of their training. They should be trained solely in metric units and decimal arithmetic. When, having acquired speed and accuracy in metric measurement, they are taught conversion to Imperial in order to cope with the remaining Imperial practice in the world outside the training school, the awkwardness of the Imperial units will be apparent to them. It is probably better to let the unit systems speak for themselves in this way rather than to attempt to persuade bilingual trainees to adopt metric practice.

CHAPTER 15

The Preparation of Changeover Programmes

15.1. Some Fundamental Considerations

Let us examine the areas of activity in which the changeover must be carried out. Every organization functions in three areas:

(a) As a purchaser of materials, goods and services offered by other organizations.
(b) As a processing agent, carrying out operations on the material purchased. This involves preparing and carrying out instructions within the organization.
(c) As a vendor, offering products to customers. This involves a knowledge of the requirements of prospective customers.

The first decision is to fix a suitable date to start the changeover. Under head (a), consider the availability of materials, information, published standards and equipment in metric language.

An attempt to go metric too far ahead of the national plan will be frustrated by delays and difficulties in obtaining materials. The draft plan prepared by the British Standards Institution shows that the beginning of 1970 is the point where sufficient British Standards in metric units, and sufficient materials, tools and technical literature should be available to enable a major effort to commence in general manufacturing industry.

Under head (b), consider the desirability or otherwise of a 'bilingual' period within the organization. If a large amount of capital plant is involved it is quite unrealistic to attempt a wholesale changeover of parts to make the entire plant metric. The changeover will extend over the life of the most recently acquired piece of plant, new items to metric standards being purchased to replace items written off, and having to coexist with remaining items to Imperial Standards.

Even in those organizations where the plant is on a scale small enough to make it feasible to consider a 'once and for all' changeover it may be necessary to have a bilingual period in the drawing office in order to build up a supply of drawings which could be used by the machine shop in its Imperial state before the changeover date, or in its metric state after the changeover.

Some organizations have carried out experiments in order to establish techniques for changing over. Some have carried out piecemeal changes; as each component or product came up for review or modification it was 'metricized' and as each machine tool amortized it was replaced by a metric successor. This method involves minimum costs but requires superb housekeeping. The prospect of mixed batches of fasteners, reamers, gauges and so on is terrible to contemplate.

Some organizations have prepared purely metric, not even bilingual, drawings and have issued them to Imperial workshops, trusting the workshop staff to carry out their own conversions.

No disasters resulting from this practice have been reported.

The fastener industry is an example of the production of metric products on Imperial machines. To the toolroom the metric dimensions merely become odd numbers of 'thous' and the product appears without any great difficulty.

Having considered the problems peculiar to the organization it is possible to consider the next important problem: customer willingness to accept metric products. In this case, as in the case of purchasing materials, there are penalties for trying to move off ahead of the majority. Until the market is ready for metric products the organization's sales force may find resistance to them. The key date for any company's products to go metric will have to be determined from a close study of the BSI plan. Naturally primary and intermediate products such as steel bar, rod and sections, fasteners, electric cables, pipes and so on, will be saleable to the assembly and finishing trades early in the national programme.

The setting up of a National Metrication Board and the passage of the necessary legislation on weights and measures for the retail trades through Parliament are matters which are the subject of recommendations in the *Report of the Standing Joint Committee on Metrication*, published in May 1968. (SBN 11 470061 3, HMSO, 2*s*. 3*d*. net.).

15.2. Drawing the Network

Consideration of the points enumerated in the above section will yield two key dates, the earliest advisable starting date and the earliest date on which the organization's metric products will be acceptable to customers. There will also be some idea of the method to be applied to the internal changeover.

The usual modern techniques of PERT and CPM are admirably suited to a problem of this type. A brief statement of the approach is not out of place here as it helps to crystallize much otherwise rather diffuse thought. PERT (Programme Evaluation and Review Technique) and CPM (Critical Path Method) both require the preparation of a network constructed from two components; a rectangle or circle to represent an event and representing a point in time, and a line linking two events and representing an activity, involving the passage of time. Events are always either the commencement or the completion of an activity.

The final event in our case will be either the marketing of a metric product or the completion of the changeover of the last item requiring adaptation.

The existence of the network, set down on paper, greatly eases the measurement of performance. By continuous monitoring of progress against the plan it is possible to detect early signs of a holdup.

It is now possible to start work on the first network, the high-level general policy plan. From this master plan we can draw detailed networks for the various functions and departments throughout the organization.

The method of network construction is a matter of individual preference, but in what follows the forward method is adopted.

To take the example of the introduction of a metric product, and to start with an announcement of company policy to offer that product:

Event 1 will be the policy announcement.
Event 2 might be the completion of the production plan.
Event 3 might be the instant at which all the materials and services for metric production are available.
Event 4 might be the offer for sale of the metric product.

Of these four key events it is number 3 which has to be most carefully matched against the national plan and against the announced intentions of the suppliers of materials and semi-finished goods.

At this point it is necessary to make the point that, to a stockholder or wholesaler, metrication brings only problems and no immediate advantages.

The last traces of Imperial stocks are not likely to disappear for some thirty to forty years, that being the life of a great deal of plant and equipment for which maintenance parts and tools will have to be stocked throughout the phasing-out period. It is not unreasonable to say that, to the stockholder and wholesaler, metrication means just another line to be carried alongside existing lines. While the long-term intention is to run down non-metric stocks the immediate prospect is one of increased capital tied up in stocks and of complicated house-keeping problems.

It is likely, therefore, that stockholders will need to be convinced of a worthwhile and increasing market for metric lines before they are willing to shoulder the burden of carrying those lines.

Between event 1 and event 2 much of the activity will be in the nature of education and training of personnel, and providing them with information on metric standards and practice. This phase is of such profound importance that it is worth examining it in some detail.

15.3. Retraining

The object of the entire exercise is to bring British practice more into line with that of the majority of mankind, and in particular with practice in continental Europe where our economic future will be decided. Merely to change from inches to millimetres achieves very little. The vital part of the exercise is to make absolutely certain that all the people involved are aware of the reasons for making the change and of the implications beyond the mere change of units. A large-scale and systematic information exercise is necessary so that everyone has the facts to ensure that the maximum benefit and the minimum cost are involved.

Many of the obstacles to the change are psychological: wrong reasons for making the change are assumed and the importance of changes in practice as well as in units is not appreciated. A little explanation early in the exercise may ease the situation.

The importance of thinking in metric units as soon as possible should be emphasized. Continual translation back into Imperial units is time wasting, exhausting, and does very little to improve facility in the use of metric quantities.

The first group to be retrained must, of course, be the design and development staff. Once a prototype metric product starts on its way through development, into production, and out onto the market a natural sequence for introducing other staff to metric practice emerges, the activity lines between events 2 and 3 and 3 and 4 will include such training.

15.4. Branches of the Network

Alongside the main route through events 1, 2, 3 and 4 there will be auxiliary routes. One might cover the preparation of sales literature on the metric product, this route could also cover sales staff training.

Another route could cover the purchasing and holding of stocks of metric materials and parts. The intricacies of dealings with sub-contractors and stockholders, as well as such matters as the acquisition of tools and gauges would properly belong in this area of the network.

Once an overall network has been drawn it is possible to draw realistic networks for the various departments within the organization, such as design office, inspection department, technical sales office and so forth. The existence of a master network will remove the risk of any one of these moving too early or too late and becoming isolated. Figure 15.1 shows a basic master network of the type described. It is reproduced from BSI *Basic Programme and Guide*, PD 6424, July 1968.

15.5. The Changeover Period

The period of phasing in metric practice and phasing out Imperial will be an ordeal. Expense and confusion will be involved. Estimates of the minimum expense range from 10% of capital invested to 28%.

The volume and variety of stocks is likely to increase. The run down of Imperial stocks is unlikely to be complete for very many years.

However, the advantages of metrication are so real that industry has itself chosen to make the change now. The penalty for delaying the change is an estimated 7% per annum rise in the cost of the operation. Few now doubt that the change is inevitable and so the most rational course seems to be to devote one's thoughts and energies to minimizing the disruption which will be caused.

15.6. Metrication in India

In 1955, on the recommendation of the Indian Standards Institution, the Government of India decided to decimalize the coinage and to adopt the metric system of weights and measures. India has now had over a decade of metrication and decimalization and the question arises of whether there are any lessons to be learnt from India's experience.

Several interesting papers have been written on the subject, both by Indian experts and by visiting British experts. The differences between the state of technology and education in India and in Britain are so great that few conclusions can be drawn. The

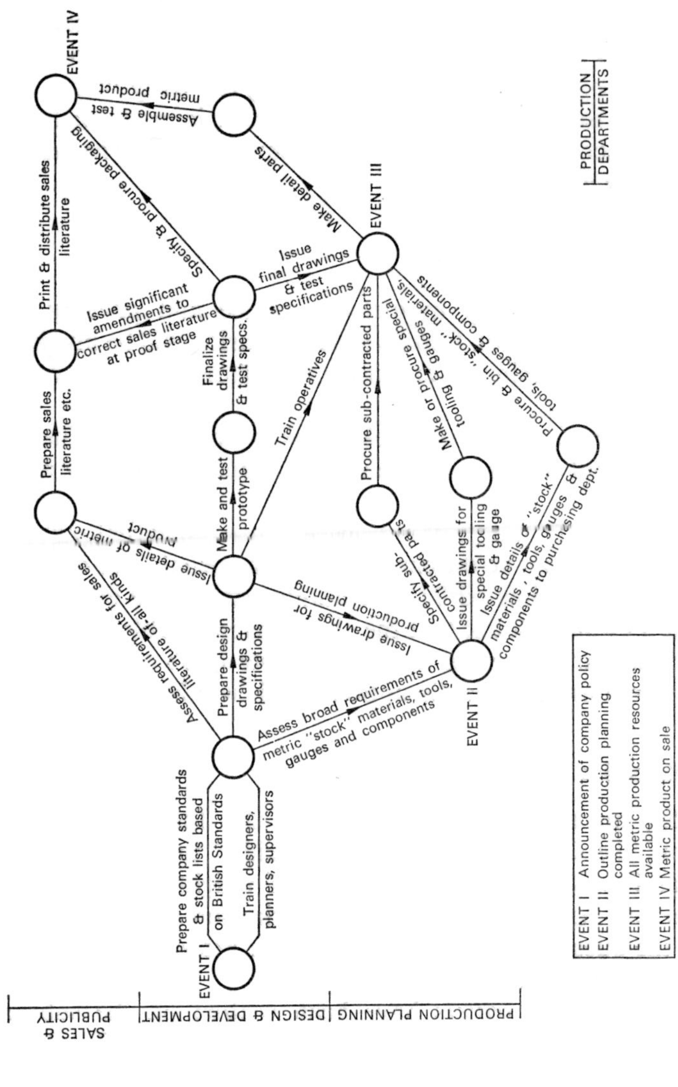

FIG. 15.1. Outline network analysis of introduction of a metric product. (Reproduced from BSI Document PD 6424, *Basic Programme and Guide*, July 1968).

Indian Government wisely decided to make the changes at a time which may fairly be said to be the dawn of their technological development. India was, and still is, a vast and diverse assembly of states, races, language and measuring systems. The introduction of the metric system by an Indian federal government is a great unifying effort. Even the British Raj failed to establish a uniform practice throughout India.

Britain, on the other hand, has over the past 200 years been developing a complex and highly industrialized society under a well-established central government and using an ancient and deeply ingrained system of currency, weights and measures.

If we bear in mind these great differences between the two situations there is no harm in setting down some of the comments that have been made.

The focus of the change was, in India, as it will be in Britain the national standards institution. The Indian Standards Institution had to solve many problems of priority. It was found, for example, that an attempt to standardize fastener sizes could not proceed until bar and rod size standards had been established.

The planning of a vast restandardization programme is a matter of the greatest difficulty because of the intricate pattern of inter-relationships which exists between stages of manufacture and between different branches of technology. The problems were complicated in India by the fact that the range of standards in existence was incomplete. The problem was not merely one of conversion and rationalization, as it is in Britain, but of establishing those standards which were missing from the list.

The ISI did valuable fundamental work on mathematical conversion, and their publications:

IS: 2–1949, *Rules for Rounding Off Numerical Values;*
IS: 786–1956, *Conversion Factors and Conversion Tables;*
IS: 787–1956, *Guide For Inter-Conversion of Values from One System of Units to Another;*
IS: 1020–1957, *Conversion Tables for Ordinary Use;*
IS: 1105–1957, *Method for Precise Conversion of Inch and Metric Dimensions to Ensure Interchangeability*

are the embodiment of much thought and work on the subject of numerical conversion. The topics discussed in Chapter 6 and Chapter 8 of this work are discussed at great length in the above documents.

Primary standardization was another subject which occupied the ISI in the early stages of the exercise. Not only did India have many indigenous systems, there were also units of the various systems bearing the same name but differing widely in value. The record stood at 100 different 'standard' values in different parts of India for what was nominally the same unit.

The language difficulty was dealt with by adhering rigidly to international terminology.

Metric units of weight, length and capacity were introduced separately with phasing-in periods of one or two years. After the transition period the metric units were declared to be the only legal ones.

One of the greatest difficulties was the shortage of trained and skilled technologists and technicians to build up an adequate weights and measures inspectorate. Millions of sets of weights, scales for linear measure, measuring vessels for volumetric standardization and conversion charts were required. Shortage of foreign exchange made it imperative to produce as much of this material as possible in India.

Much scepticism and resistance was met with at first but this was soon replaced by realization that the administration intended to be firm in the matter and the view that what had to be done had better be done quickly prevailed. An élite corps of senior inspectors, numbering about 1,000, set about the task of preventing the new legislation from becoming a dead letter in the statute book. A massive publicity campaign prevented the public from pleading lack of information. Posters and demonstrations drove home the new units, and their names, values and the benefits consequent upon their adoption.

The Indian Railways published new timetables in the new decimal currency, with distances in kilometres. Baggage and freight weighing-machines were recalibrated in kilogrammes.

Petrol pumps were recalibrated. Conversion parts were, where-ever possible, made in India. The cost was about 1% of the first pessimistic estimates.

The attitude in India today seems to be that the nation has gone a long way towards standardization but that much remains to be done. The first decade was the worst period and a new generation of school leavers, trained in metric and decimal units from the first, will be the first to feel the advantages of the change. The consensus of opinion on cost estimates is that these are usually much too high and are misleading. The importance of machine-tool conversion is felt to have been over-emphasized at first. An inch-calibrated machine tool can produce metric components quite successfully for the remainder of its useful life. It is felt that the most important part of the whole exercise was adequate retraining courses. Progress in education and training was hampered at first because most textbooks were of English or American origin. There is determination to set this matter right to avoid burdening the next generation with the chore of conversion.

Industry will only be truly metric when the next generation of products designed and constructed wholly in the metric mode, appears.

Government pressure is being maintained to ensure that multilingualism in units does not become a habit.

Bibliography

Chapters 1 and 2

Encyclopaedia Britannica: Articles on Metric System, Metrology and Physical Units. Various editions.

Chapter 3

BS 3763: 1964, *The International System (SI) Units.*
BS 5686: 1967, *The Use of SI Units.*
BS 2990: 1958, *Rationalized and Unrationalized Formulae in Electrical Engineering.* (With Amendment Jan. 1960).
ANDERTON, P. and BIGG, P. H., *Changing to the Metric System*, HMSO, 1967.
KAYE and LABY, *Tables of Physical and Chemical Constants*, Longmans, various editions.

Chapter 4

ISO Memento, 1967.
ISO Catalogue, 1967.
BSI News, published monthly.
BSI Yearbook and Supplement, 1967.

Chapter 5

BS 2045: 1965, *Preferred Numbers.*
ISO R 286: 1962, *Limits and Fits.*
BS 1916: 1953, Part 2: *Guide to the Selection of Fits.*
BS 3643: 1966, *ISO Metric Screw Threads.*
BS Handbook 18: 1966, *Metric Standards for Engineering.*
BS 4318: 1968, *Preferred Metric Basic Sizes for Engineering.*

Chapter 6

BS 350: 1962, *Conversion Factors and Tables.*
BS 1957: 1953, *Presentation of Numerical Values, Rounding of Numbers.*
BS 2856: 1957, *Precise Conversion of Inch and Metric Sizes on Engineering Drawings.*
BS Handbook 18: 1966, *Metric Standards for Engineering.*
ISO R 286: 1962, *Limits and Fits.*

Chapter 7

BS 308: 1964, *Engineering Drawing Practice.*

ISO R 128: 1959, *Engineering Drawing, Principles of Presentation.*

ISO R 129: 1959, *Engineering Drawing, Dimensioning.*

CONWAY, H. G., *Engineering Tolerances*, Pitman, 3rd edition, 1966.

ABBOTT, W., *The Dimensioning of Engineering Drawings*, Blackie, 1953.

PARE, FRANCIS and KIMBRELL, *Engineering Design*, Holt, Rinehart & Winston, 1963.

Chapter 8

ISO R 286: 1962, *Limits and Fits.*

BS 1916: 1953, *Limits and Fits for Engineering.*

CONWAY, H. G., *Engineering Tolerances*, Pitman, 3rd edition, 1966.

PARE, FRANCIS and KIMBRELL, *Engineering Design*, Holt, Rinehart & Winston, 1963.

BS Handbook 18: 1966, *Metric Standards for Engineering.*

Chapter 9

MACHINERY, *Machinery's Screw Thread Book*, 19th edition, 1965.

BS 3580: 1964, *The Strength of Screw Threads.*

BS 3643: 1966, *ISO Metric Screw Threads.*

BS 3692: 1967, *ISO Metric Precision Hexagon Bolts, Screws and Nuts.*

BS 4190: 1967, *ISO Metric Black Bolts, Screws and Nuts.*

BS Handbook 18: 1966, *Metric Standards for Engineering.*

Supplement No. 1: 1968 to BS 192: 1961, *Open Ended Spanners.*

Supplement No. 1: 1968 to BS 2558: 1954, *Tubular Box Spanners.*

Supplement No. 1: 1968 to BS 3555: 1962, *Ring Spanners.*

Supplement No. 1: 1968 to BS 4006: 1966, *Socket Spanners.*

Chapter 10

DECIMAL CURRENCY BOARD, *Decimal Currency: Expression of Amounts in Printing, Writing and Speech*, HMSO, 1968.

DECIMAL CURRENCY BOARD, *Decimal Currency: Three Years to Go*, HMSO, 1968.

Financial Times.

Whitakers Almanack, 1968.

BOARD OF TRADE, *Guides to Businessmen.*

Chapter 11

SLIWA, A. J. and FAIRWEATHER, L., AJ Metric Handbook, *Architects Journal*, 1968.

ELDER, A. J., AJ Metric Guide to the Building Regulations, *Architects Journal*, 1968.

MINISTRY OF PUBLIC BUILDING AND WORKS, *Going Metric in the Construction Industry*, HMSO, 1967.

ROYAL INSTITUTION OF CHARTERED SURVEYORS, *Metric Guide*, RICS, March 1968.

PD 6030: 1967, *Programme for the Change to the Metric System in the Construction Industry*, BSI.

PD 6031: 1967, *A Guide for the Use of the Metric System in the Construction Industry*, BSI.

BS 4011: 1966, *Basic Sizes for Building Components*.

Chapter 12

BS 1637: 1950, *Memorandum on the MKS System of Electrical and Magnetic Units*. (Decisions taken by the International Electrotechnical Commission.)

BS 3979: 1966, *Dimensions of Electric Motors (Metric Series)*.

BRITISH INSULATED CALLENDER'S CABLES LTD., *Electrical Conductors, International Standard Sizes*, Publication No. 377, BICC, 1957.

ILLUMINATING ENGINEERING SOCIETY, *Code for the Interior Lighting of Buildings*, IES, 6th edition.

BS 3820: 1964, *Electric Lighting Fittings*.

CENTRAL ELECTRICITY GENERATING BOARD, *An Introduction to the Metric System*, CEGB, 1967.

Chapter 13

BS 4318: 1968, *Preferred Metric Basic Sizes for Engineering*.

BS Handbook 18: 1966, *Metric Standards for Engineering*.

ENGINEERING MATERIALS AND DESIGN, *Metrication Reference*, E.M.&D.,1968.

ISO R 54: 1966, *Modules and Diametral Pitches of Cylindrical Gears for General Engineering*.

BS 21: 1957, *Pipe Threads*.

BS 870: 1950, *External Micrometers*.

BS 887: 1950, *Vernier Callipers*.

CONWAY, H. G., *Engineering Tolerances*, Pitman, 3rd edition, 1966.

Chapter 14

Publications by the Department of Education and Science:

Circular Letter F.E.C.L. 4/68 (T.23/56/01), March 1968.

Introduction of the Metric System, Policy Statement No. 1, Council of Technical Examining Bodies.

Memorandum No. 1 (Construction Industry). A Background Note on changing to the metric system.

Chapter 15

PD 6424: 1968, *The Adoption of the Metric System in Engineering: Basic Programme and Guide*, BSI, July 1968.

INSTITUTION OF PRODUCTION ENGINEERS, *Metrication in the United Kingdom*, I. Prod. E., 1967.

SLIWA, A. J. and FAIRWEATHER, L., AJ Metric Handbook, *Architects Journal*, 1968.

ENGINEERING EQUIPMENT USERS ASSOCIATION, *Change to the Metric System*, EEUA, 1967.

ENGINEERING MATERIALS AND DESIGN, *Metrication Reference*, E.M.&D. 1968.

COVENTRY CITY LIBRARIES, *Metrication — Decimalisation, a Bibliography*, September 1967.

STIRES and MURPHY, *Program Evaluation and Review Technique and Critical Path Method*, Materials Management Institute, 1963.

Note

British Standards, PD Documents, ISO Recommendations, and publications by the IEC and CEE, may be obtained from:

British Standards Institution,
Sales Branch,
Newton House,
101–113 Pentonville Road,
London N.1.

Appendix 1

Conversion Factors

A COMPARISON table of the various unit systems in common use is provided, together with conversion factors from customary British units into SI.

Table App. 1.1. Comparison Table of Unit Systems in Common Use.
Table App. 1.2. Factors for Conversion to Units of SI.
Table App. 1.3. Prefixes and Multipliers.
Table App. 1.4. Combined Units; Factors for Conversion to Units of SI and other Metric Units.

TABLE APP. 1.1. COMPARISON TABLE OF UNIT SYSTEMS IN COMMON USE

Dimension	Imperial foot-pound-second system	Imperial Engineering system	Metric (cgs) System	Metric Technical (Engineering) System	System International	Metre-Tonne-second system
Length	foot (ft)	foot (ft)	centimetre (cm)	metre (m)	metre (m)	metre (m)
Mass	pound (lb)	slug (lb ft^{-1} sec^2)	gramme (g)	mug ($kg^{-1}sec^2$)	kilogramme (kg)	tonne (10^3kg)
Time	second (sec)	second (sec)	second (sec)	second (sec)	second (s)	second (sec)
Force	poundal (lb ft sec^{-2})	pound (lbf) or (lb)	dyne (g cm sec^{-2})	kilogramme (kgf) or (kg)	newton (kg m sec^{-2})	sthene (sn) (tonne m sec^{-2})
Used by	a few textbook writers	Engineers in America, Britain, Canada, Australia	Physicists & Electrical Engineers, everywhere.	Engineers outside U.S. & U.K.	Intended to become universal.	Some engineers in the metric world
Also used:		ton force (2240 lb f) (ton f) kip force (1000 lb f) (kip f)				

TABLE APP. 1.2. FACTORS FOR CONVERSION TO UNITS OF SI

Present unit	Unit of SI	To convert from present unit to unit of SI multiply by:
Acre	Square metre (m^2)	4,046.9
Angstrom (Å)	Metre (m)	10^{-10}
Atmosphere (atm)	N/m^2	101,325
Bar (b)	N/m^2	10^5
British thermal unit (B.t.u.)	Joule (J)	1,055.06
Calorie (cal)	Joule (J)	4.186 8
Cm of Hg	N/m^2	1,333.2
Cm of water	N/m^2	98.066 5
Dyne (dyn)	Newton (N)	10^{-5}
Erg	Joule (J)	10^{-7}
Foot (ft)	Metre (m)	0.304 8
Gallon (gal)	Cubic metre (m^3)	$4.546\,1 \times 10^{-3}$
Gauss	Wb/m^2	10^{-4}
Grain (gr)	Kilogramme (kg)	$6.479\,9 \times 10^{-5}$
Horsepower (hp)	Watt (W)	745.70
Hour (h)	Second (s)	3,600
Hundredweight force (cwtf)	Newton (N)	498.20
Hundredweight mass (cwt)	Kilogramme (kg)	50.802
Inch (in.)	Metre (m)	0.025 4
Inch of Hg	N/m^2	3,386.4
Inch of water	N/m^2	249.09
Kilocalorie (kcal)	Joule (J)	4,186.8
Kilogramme force (kgf) Kilopond (kp)	Newton (N)	9.806 65
Litre (l)	Cubic metre (m^3)	0.001
Maxwell	Weber (Wb)	10^{-8}
Micron (μm)	Metre (m)	10^{-6}
Mile	Metre (m)	1,609.3
Oersted	A/m	79.577
Ounce fluid (fl oz)	Cubic metre (m^3)	$2.841\,3 \times 10^{-5}$
Ounce force (ozf)	Newton (N)	0.278 01
Ounce mass (oz)	Kilogramme (kg)	0.028 350
Pint (pt)	Cubic metre (m^3)	$5.682\,6 \times 10^{-4}$
Poise (P)	Ns/m^2	0.1
Pound force (lbf)	Newton (N)	4.448 2
Pound mass (lb)	Kilogramme (kg)	0.453 59
Stokes (St)	m^2/s	10^{-4}
Therm	Megajoule (MJ)	105.51
Ton force (tonf)	Newton (N)	9,964.0
Ton mass (ton)	Kilogramme (kg)	1,016.0
Torr	N/m^2	133.32
Yard (yd)	Metre (m)	0.914 4

TABLE APP. 1.3. PREFIXES AND MULTIPLIERS

Multiples and submultiples of the units are formed by means of the following prefixes:

Prefix	Symbol	Multiple
tera	T	10^{12}
giga	G	10^{9}
mega	M	10^{6}
kilo	k	10^{3}
hecto	h	10^{2}
deca	da	10^{1}
deci	d	10^{-1}
centi	c	10^{-2}
milli	m	10^{-3}
micro	μ	10^{-6}
nano	n	10^{-9}
pico	p	10^{-12}
femto	f	10^{-15}
atto	a	10^{-18}

TABLE APP. 1.4. COMBINED UNITS; FACTORS FOR CONVERSION TO UNITS OF SI AND OTHER METRIC UNITS

	Present unit	SI unit	To convert from present unit to SI unit multiply by:
Area	in^2	m^2	$6.451\ 6 \times 10^{-4}$
	ft^2	m^2	0.092 903
	yd^2	m^2	0.836 13
Volume	in^3	m^3	$1.638\ 7 \times 10^{-5}$
	ft^3	m^3	0.028 317
	yd^3	m^3	0.764 55
Velocity	ft/s	m/s	0.304 8
	mile/h	m/s	0.447 04
Acceleration	ft/s^2	m/s^2	0.304 8
Density	lb/in^3	kg/m^3	27.680
	lb/ft^3	kg/m^3	16.018
	oz/pt	kg/m^3	49.888
	lb/gal	kg/m^3	99.776
Flow	ft^3/min	m^3s	$4.719\ 5 \times 10^{-4}$
	ft^3/s	m^3/s	0.028 317
	gal/min	m^3/s	$7.576\ 8 \times 10^{-5}$
	gal/s	m^3/s	$4.546\ 1 \times 10^{-3}$
Pressure, stress	dyn/cm^2	N/m^2	0.1
(see Chapter 3, §7, on hectobar)	kgf/cm^2 kp/cm^2	N/m^2	98,066.5
	lbf/in^2	N/m^2	6,894.8
	lbf/in^2	MN/m^2*	$6.894\ 8 \times 10^{-3}$
	lbf/ft^2	N/m^2	47.880
	$tonf/in^2$	MN/m^2*	15.444
	$tonf/ft^2$	MN/m^2*	0.107 25
Work, energy	ft lbf	J	1.355 8
	in. ozf	J	$7.061\ 5 \times 10^{-3}$
	hp h	MJ	2.684 5
	Wh	J	3,600
	kWh	MJ	3.6
	litre atm	J	101.33
Power	B.t.u./h	W	0.293 07
	erg/s	W	10^{-7}
	ft lbf/min	W	0.022 597
	ft lbf/s	W	1.355 8
Torque, Moment	lbf ft	Nm	1.355 8
	lbf in.	Nm	0.112 98

	Present unit	SI unit	To convert from present unit to SI unit multiply by:
Thermal capacity	B.t.u./lb °F	J/kg°K	4,186.8
Specific heat, Heat Equivalent, Latent heat	cal/g °C	J/kg°K	4,186.8
	B.t.u./lb	J/kg	2,362
	cal/g	J/kg	4,186.8
Calorific value (volumetric)	B.t.u./ft³	J/m³	37,259
	B.t.u./gal	J/m³	232,080
	cal/cm³	J/m³	$4.186\,8 \times 10^{6}$
Thermal conductivity	B.t.u. in./ft²h°F	W/m K	0.144 23
	cal/cm s°C	W/m K	418.68
Thermal diffusivity	in. ft/h	m²/s	$2.150\,5 \times 10^{-6}$
Surface tension	dyn/cm	N/m	0.001
Permeability of porous solid to specific fluid	cm⁴/s dyn	m⁴/Ns	10^{-3}
	cm²/s cm Hg	m⁴/Ns	$7.506\,4 \times 10^{-8}$
	cm²/s cm water	m⁴/Ns	$1.019\,7 \times 10^{-6}$
Absolute permeability of porous solid	cm² cP/s atm, Darcy	m²	$9.869\,2 \times 10^{-13}$
	cm⁴P/s dyn, Perm, cm²	m²	10^{-4}
	cm²P/s cm Hg	m²	$7.506\,4 \times 10^{-9}$
	cm²P/s cm water	m²	$1.019\,7 \times 10^{-7}$

*The meganewton per square metre (MN/m²) which is identical with the newton per square millimetre (N/mm²) is recommended for strength and elasticity values, see also Chapter 3, §7, on the hectobar.

Electrical Units

The volt (V), ampere (A), ohm (Ω), coulomb (C), watt (W), henry (H) and farad (F) have been incorporated into the SI system.

Appendix 2

Published Programmes for Metrication

Programmes prepared by:

British Standards Institution	Fig. App. 2.1
Institution of Production Engineers	Fig. App. 2.2
Construction Industry	Fig. App. 2.3
Central Electricity Generating Board	Fig. App. 2.4
Engineering Equipment Users Association	Fig. App. 2.5

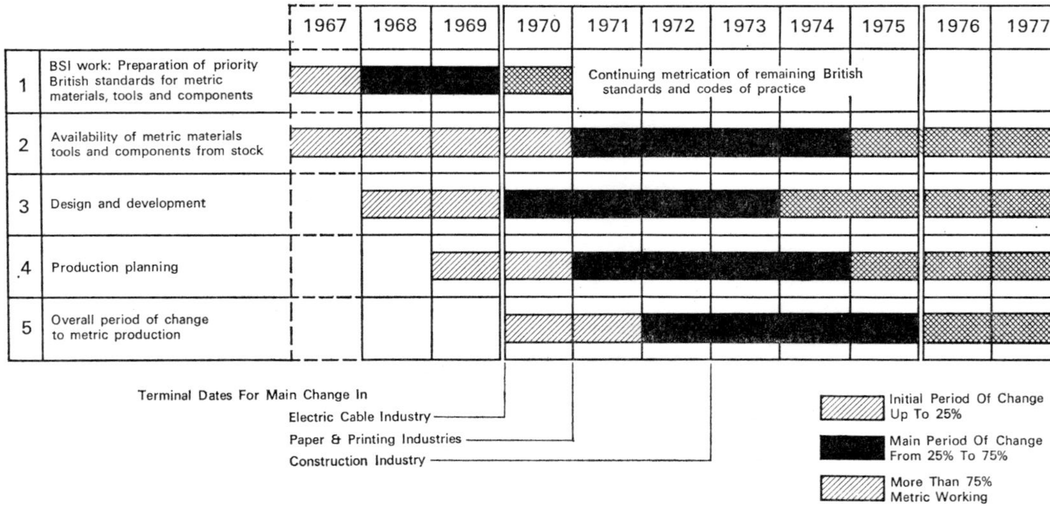

FIG. APP. 2.1. The Adoption of the Metric System in Engineering: Basic Programme and guide, PD 6424. Published by BSI in July 1968.

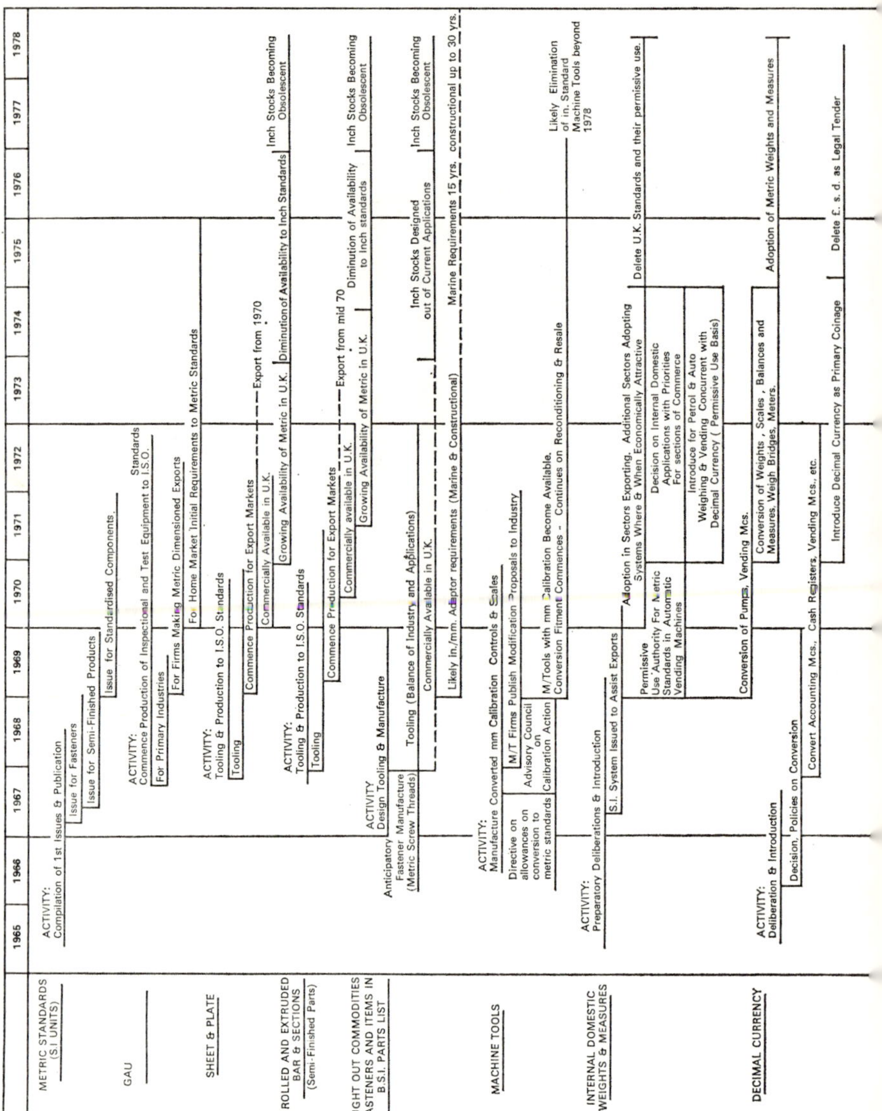

FIG. APP. 2.2. Application of the Metric System to Industry and Commerce. Published by the Institution of Production Engineers in January 1967.

Index

Adoption of metric system in engineering 4
Alternative conversion procedure 54
Ampere, definition of 18
Antiquarian drawing size 58
Architect's Journal 109
Associability chart 124

Brick, the metric 107
British Standards Institution 3
British Thermal Unit 34
Building Regulations 109

Caesium, transition 8
Calorie 33
Camera alignment marks 63
Canadian Parliament 14
Candela 9
Celsius, degree interval 18
Centimetre, gramme second system 8
Central Electricity Generating Board 113
Characteristic impedance of free space 31
Circuits
  printed 3
  printed, standards for 112
Class of fit 68, 82
Coarse rounding 55
Coherence 19
Compatibility, international 2
Conduit, electrical 111
Confederation of British Industry 3
Conference, general, of weights and measures 5
Congress, U.S. 14
Construction industry programme 108
Convention, International Metric 9
Conversion
  accuracy 70
  of traditional to decimal coinage 93
  precise 50
Cost of metrication 131
Critical path method 128
Curie 18
Currency, decimal 89

Decimal structure 1
Design form of thread 81
Deviations 66
Diametral pitch 116
Dimensional analysis 25
Drawing
  for microfilming 62
  sheet sizes 57–59
Dynamics 20

Einstein 16
Electric space constant 31
Electrical S.I. units 28

Electromagnetic units 8
Electrostatic units 8
Electrotechnical, International, Commission 8, 41
Euronorm 37
Exchange rates, international 91, 96–105
Exports, British 2

Fineness of rounding 52
Force, unit of 19
Fracture toughness testing 23
French Revolution 5
Functional interchangeability 51

Gauge tolerances 56
Gauss 28
Gears, I.S.O. modular system 116
Geodetic definition of metre 5
GIORGI, PROFESSOR 8, 31
Giorgi, rationalized M.K.S.A. units 10
Gravity 20

Hamburg drawing board 57
HEAVISIDE, OLIVER 31
Hectobar 22
Hertz 22
High-tensile fasteners 84
Hodgson Committee 3
Hole basis system 69

India, metrication in 131
Indian Standards Institution 133
Industrial Research Incorporated 14
Instruments, measuring 119
Interference between nut and bolt 83
International Electrotechnical Commission 8, 41
International Standards 37
International Standards Organization 37
Izod testing 115

Joule 21, 33
Joules mechanical equivalent 34

Kelvin 8
  degree interval 18
Kilogramme force 20
Kilogramme, international prototype 6

Lighting, interior, of buildings 111
Limits and fits 48
  I.S.O. system 64
Litre, definition of 16
Luminous intensity 9
Lux 111

Magnetic, space constant 9
Maxwell 28
MAXWELL, JAMES CLERK 32
Mètre des Archives 5
Metric technical units 19
MICHELSON, A. A. 5
Microfilming of drawings 61
Micrometers 119
Military specifications, U.S. 112
Modular sizes 106
Modular Society 108
Module 3
Mol 19
Multiples, variety in Imperial System 1

National Standards 35
Network
  branches 130
  programming 128

Newton
definition of 19
second law 19
Newtonian fluids 27
Nominal designations 51
Nuts, suitable, for high-tensile bolts 86

Parliament 1
Pipe threads 117
Pipes, large-diameter 3
Platinum–iridium
kilogramme 6
prototype metre 5
Poise 27
Pound, Imperial Standard 6, 7
Pound sterling 89
Power 21
Precise conversion 50
Preferred numbers 42
Prefixes 15
and multipliers 15
Presentation of toleranced dimensions 60
Primary standards of mass 6, 7
Programme evaluation and review technique 128
Projection conventions 57
Proof-load stress 87
Pseudo-metrication 2

Quadrant, Earth's Polar 5

Radian 18
Rationalization
of electrical units 9
of screw threads 76
RENARD, COLONEL CHARLES 42
Renard Series 42
Reyn 27
Rounding of numbers 52
Rules for use of prefixes 17

Scales for drawings 60
Schools Council 121
Screw thread, I.S.O. metric 77
Screw threads 48
Second
atomic 8
ephemeris 7
universal 7
Shaft basis system 69
Sizes, preferred 115
Specific heat 32
Standards, International, Organization 2
Statics 20
Stein 16
Steradian 18
Stocks
variety of 2
volume of 2
Stokes 28
Strength of screw threads 79
Subgeometric progression 46
Supplementary units 18
Surveyors, Chartered, Royal Institution of 109
System, International 15

Table calorie 34
Tapers
5% metric 118
Morse 118
Tensile testing of steel 114
Text books 123
Tolerance 66
Toleranced dimensions, presentation of 60
Torque 21
Training, specialized 122

Unified threads 76
United States of America 9, 13
Units, basic S.I. 15

Velocity, angular 22
Vernier gauges 119
Viscometer
  Engler 28
  Redwood 28
  Saybolt 28
Viscosity, units of 25
Visual aids 124

Watt 21
Wavelength
  of cadmium light 5
  of radiation of krypton-86 6
Weighing, double 6
Weight 19
Whitworth threads 76
Work 20

Yard, Imperial Standard 6
Young's Modulus 24